The Shadow of Enlightenment

The Shadow of Enlightenment

Optical and Political Transparency in France, 1789–1848

Theresa Levitt

OXFORD
UNIVERSITY PRESS

OXFORD

UNIVERSITY PRESS

Great Clarendon Street, Oxford OX2 6DP

Oxford University Press is a department of the University of Oxford.
It furthers the University's objective of excellence in research, scholarship,
and education by publishing worldwide in

Oxford New York

Auckland Cape Town Dar es Salaam Hong Kong Karachi
Kuala Lumpur Madrid Melbourne Mexico City Nairobi
New Delhi Shanghai Taipei Toronto

With offices in

Argentina Austria Brazil Chile Czech Republic France Greece
Guatemala Hungary Italy Japan Poland Portugal Singapore
South Korea Switzerland Thailand Turkey Ukraine Vietnam

Oxford is a registered trade mark of Oxford University Press
in the UK and in certain other countries

Published in the United States
by Oxford University Press Inc., New York

British Library Cataloguing in Publication Data
Data available

Library of Congress Cataloging in Publication Data
Levitt, Theresa.
The shadow of enlightenment : optical and political transparency
in France, 1789–1848 / Theresa Levitt.
p. cm.
ISBN 978–0–19–954470–7
1. Optical engineering—France—History—18th century.   2. Optical
engineering—France—History—19th century.   3. France—Politics and
government—18th century.   4. France—Politics and government—
19th century.   5. Enlightenment.   I. Title.
TA1522.L48   2009
530.0944'09033—dc22   2008048422

Typeset by Newgen Imaging Systems (P) Ltd., Chennai, India
Printed on acid-free paper by the
MPG Books Group
in the UK

ISBN 978–0–19–954470–7 (Hbk.)

1 3 5 7 9 10 8 6 4 2

Acknowledgements

I have been continually delighted by the generosity of the many people who have made this work possible. Their insights and amendments have made this book immeasurably better. Peter Galison, Lorraine Daston, and Mario Biagioli saw the project through its early stages and expanded its possibilities. Jed Buchwald offered valuable guidance on several sections. Many friends, colleagues and mentors have contributed through comments and conversations, notably David Aubin, David Barnes, Charlotte Bigg, Robert Brain, Jimena Canales, Deborah Coen, Olivier Darrigol, Michael Gordin, Matt Jones, David Kaiser, John Krige, Bruno Latour, Michael Lynch, Sharrona Pearl, Denise Phillips, Jessica Riskin, Simon Schaffer, Sam Schweber, Rena Selya, Grace Shen, Joel Snyder, Matt Stanley, Mary Terrall, John Tresch, Debbie Weinstein, David Wilson, and Norton Wise. My colleagues at the University of Mississippi have been immensely supportive. I want to thank Nancy Bercaw, Erin Chapman, Sue Grayzel, Angela Hornsby-Gutting, Marc Lerner, Annette Trefzer, Joe Ward, and Noell Wilson for their help.

I am very appreciative of the institutional support of both Harvard and the University of Mississippi, as well as the Fulbright Program, the National Science Foundation, and the Max Planck Institute for the History of Science. I also want to thank the staff of the libraries and archives that provided the material for this work, in particular the archives of the Academy of Sciences, the Bibliothèque de l'Institut de France, the Bibliothèque de l'Observatoire, the Maison Auguste Comte, the French National Archives, the Bibliothèque Nationale, the National Museum of Photography, Film, and Television, and the Royal Society. Further gratitude goes to Sönke Adlung and his colleagues at the Oxford University Press for all their attention to the manuscript.

I dedicate this work to my parents, who made it all possible, and Björn and Juneau, who made it all worthwhile.

Table of Contents

The Shadow of Enlightenment: Optical and Political Transparency in France, 1789–1848

Hé, monsieur, un roman est un miroir qui se promène sur une grande route. Tantôt il reflète à vos yeux l'azur des cieux, tantôt la fange des bourbiers de la route.[1]

Stendhal, *Le rouge et le noir*

"A novel is a mirror," Stendhal famously wrote in 1830, and with the statement launched realism's most enduring icon: the silvered surface reflecting unapologetically the sky or mud found before it. Stendhal's aphorism seems to hint at the possibility of transparent reflexivity, pure reflection where the representation of the world was a perfect match for the world itself. Whether Stendhal was championed as uncompromising or upbraided as naive, readers have generally assumed that a mirror reflected everything in front of it.

But the mirror was no longer so innocent. Only a few years before, astronomers at the Paris Observatory had discovered that light, when reflected off a mirror, underwent a subtle alteration. It seemed to line itself up in a single direction, instead of the haphazard state it was normally in. The phenomenon, known as polarization, had been known for centuries as the obscure consequence of certain doubly refracting crystals. But with the discovery that light could become polarized through reflection as well, everything changed. Now all light possessed what had previously been viewed as a rare quality. This raised the question: what exactly does a mirror reveal? The light that reflected off may have looked the same, but it was altered. An alteration imperceptible to the eye, but easy enough to reveal with a little optical skill.

The polarization of light may seem like a recondite bit of scientific minutiae, certainly far from the mind of Stendhal as he penned his brief theory of literature. But that is not necessarily the case. As an enthusiastic follower of the physical sciences, Stendhal could not have been unaware of how complicated mirrors had become. He wrote in particular about the work of two young physicists, François Arago and Jean-Baptiste Biot, who had materialized the

---

[1] "A novel is a mirror that strolls along a highway. Now it reflects the blue of the skies, now the mud puddles underfoot." Stendhal, *Le rouge et le noir* (Paris: Gallimard, 2000), 479.

strange phenomenon into an imaging device, the polarimeter. These devices, representing the latest in optical technology, had a mirror at one or both ends. Light would reflect off the first of the mirrors, and emerge from the other end as a splash of brilliant complementary colors. When attached to a projection apparatus, the instrument displayed flowers, birds, and other colorful images. Arago began touring the salons with his device, including those attended by his close acquaintance Stendhal (Figure 0.1).

The mirror had lost its status as perfect reflector in an age that called the possibility of transparency into question. The Enlightenment project of the eighteenth century had in large part been built around the project of expanding visibility. The world was seen as eminently legible, if only one were looking correctly. Optical instruments were recruited into the project of visibility. Previously, telescopes and microscopes had magnified the world, bringing it closer but leaving it essentially the same (this was of course part of a fiction crucial for the instruments' success). With the polarimeter, however, the rules changed. Instead of rendering visible things that were too far away or too small to be seen, it gave form to precisely what the naked eye could not see. The polarimeter, an instrument of the nineteenth century, had to make its way in a landscape where the lines of visibility were being redefined. Even Arago and Biot, who had invented it, could not agree on what their instruments were showing them.

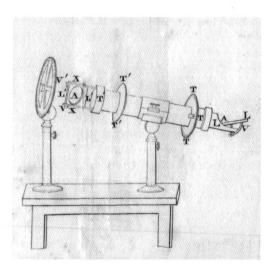

Fig. 0.1 Arago's polarimeter. The mirrors are represented by the letter A. The light hits the first mirror on the right-hand side, and reflects into the tube. It then passes through a thin sheet of mica or mica-like substance inside the tube, then emerges and is reflected once again off the mirror on the left-hand side. (From E.G. Fischer, *Physique mécanique*, J. B. Biot, trans., 4th ed. (Paris: Bachelier, 1830) pl. 12.)

This book traces the newly ambiguous status of transparency through the careers of Arago and Biot. The two men were young and little known when they each presented their version of a polarimeter, but over the next forty years they emerged as major figures in the physics community. Their nominal friendship in 1811 gave way to the bitterest of rivalries. Their disagreement began on the subject of polarized light, but expanded to encompass nearly every aspect of their lives. Politics, religion, agricultural policy, education, dinner companions, housing arrangements, photography, railroads, vital forces, astrology, and the Egyptian calendar formed the grist of the two rivals' own polarized discourse.

The optical revolution

Classic histories of science have long recognized the development of optics in France in the first decades of the nineteenth century as a seminal moment in the creation of a rigorous mathematical physics. Thomas Kuhn used the episode as his first example of a revolution in science in *The Structure of Scientific Revolutions*.[2] Arago and Biot's work on polarization was at the heart of an "optical revolution," transforming scientists' relation to the natural world. The "new optics" combined quantitative mathematical analysis with experimental precision, and marked the transition from the natural philosophy of the eighteenth century to the distinct field of physics.[3] Originating in France, the optical revolution swept across Europe, transforming not only theories of light, but also instrumental practices and modes of explanation.[4] In the mathematical physics seminars of German Universities and the Analytical Society of Cambridge, French optical work, and particularly theories of polarization, formed the template for the emerging discipline of physics.[5] Theories of light were thus at the center of a total reorganization of how one studied the world.

Yet discussions of light could hardly be contained to mathematical physics seminars alone. Scarcely was there a time or place more awash in the discourse of light than Paris in the nineteenth century. The City of Light still fashioned itself as the center of the project of *Lumières*. "Illuminer," "rayonner," "éclairer," the very language put light at the center of a cultural project that

---

[2] Thomas Kuhn, *The Structure of Scientific Revolutions* (Chicago: University of Chicago Press, 1970), 11–12.

[3] Jed Buchwald, *The Rise of the Wave Theory of Light: Optical Theory and Experiment in the Early Nineteenth Century* (Chicago: University of Chicago Press, 1989).

[4] Xiang Chen, *Instrumental Traditions and Theories of Light: The Uses of Instruments in the Optical Revolution* (Dordrecht: Kluwer Academic Publishers, 2000).

[5] Christa Jungnickel and Russell McCormmach, *Intellectual Mastery of Nature: Theoretical Physics from Ohm to Einstein* (Chicago: University of Chicago Press, 1986), 44, 85; Kathryn Olesko, *Physics as a Calling: Discipline and Practice in the Konisgberg Seminar for Physics* (Ithaca: Cornell University Press, 1991); J.M. Dubbey, "The Introduction of the Differential notation into Great Britain," *Annals of Science*, 19 (1963): 37–48; Geoffrey Cantor, *Optics After Newton: Theories of Light in Britain and Ireland, 1704–1840* (Manchester: Manchester University Press, 1983), 148.

was still picking up steam in the nineteenth century. Light was the medium through which one learned about the world. At the heart of Western tradition was a profound commitment to what Martin Jay has called ocularcentrism.[6]

In the first years of the nineteenth century, the symbolic power of light took on a specifically modern configuration. The "optical revolution" was also a fitting title for the profound changes attendant in the creation of a modern visual culture. Gone was the firm security in light that guaranteed a connection between subject and object. Opacity emerged to take its place alongside illumination. For Hans Blumenberg, the use of limelight in theaters captured the nineteenth century's new metaphorics, in which darkness was the natural state and lighting became a technology for directing vision.[7] With the dream of complete transparency gone, visibility became a form of discipline. No longer was there the hope that *every*one would see *every*thing. Rather, strict control determined who saw what. Foucault chose the panopticon as the symbol of the age not, as the name might imply, because it represented the triumph of universal vision. Rather, the central viewing tower encircled by prison cells bespoke a careful arrangement of the visible and invisible, where the prisoner was seen by all but saw nothing himself.[8] Visibility was inseparable from issues of control and authority, seated at the nexus between power and knowledge.

The modern subject was a particularly ocular being, born in this light and defined by the ability to see. Jonathan Crary has traced the creation in the 1820s and 1830s, of a new kind of observer, endowed with a thick subjectivity by an emerging constellation of social forces and institutions designed to act on the viewing body. Key to the development of this modern observer was a transformation in scopic regime, where the camera obscura of the seventeenth and eighteenth centuries, intent on distancing the object and observer, gave way to the more corporeally implicated optical devices of the nineteenth century.[9] The "optical revolution" was not just an internal event in the discipline of physics, but could be used to describe the birth of a modern visual culture.

The practice of representation

Light was an inescapable metaphor, saturating the age. But light was also more than a metaphor for knowledge. It was, quite literally, the primary way that people learned about the world. It is this literal aspect of light's illuminating power that forms the subject of this book. What access did light allow one to the world? How did light function as an arbiter of visibility? Optics, as the

[6] Martin Jay, *Downcast Eyes: The Denigration of Vision in Twentieth-Century French Thought* (Berkeley: University of California Press, 1993).

[7] Hans Blumenberg, "Light as a Metaphor for Truth: At the Preliminary Stage of Philosophical Concept Formation," Joel Anderson, trans. in *Modernity and the Hegemony of Vision*, David Michael Levin, ed. (Berkeley: University of California Press, 1993).

[8] Michel Foucault, *Surveiller et Punir* (Paris: Gallimard, 1993).

[9] Jonathan Crary, *Techniques of the Observer: On Vision and Modernity in the 19th Century* (Cambridge, MA: MIT Press, 1992).

scientific study of light, addressed precisely these concerns. Optical instruments were at the center of the question of how things were made visible. Their practitioners were very concerned with establishing the connection between what their instruments showed and what existed in the world. Much of the efforts of the men (and, sadly, in this particular story, men only) engaged in the study of optics was spent trying to connect whatever emerged from their optical devices back to the world. Enormous thought and energy went into assigning meaning to the products of optical devices. The manipulations they performed upon light as it passed through their lenses and colored glass were intended to reveal the hidden properties of the light that entered their instrument. Ultimately, their actions were a particularly painstaking act of representation, tying together the world and its appearance.

The study of representation has proven enormously valuable in the domains of both knowledge and politics. It is a curiously malleable notion, however, whose very meaning seems to change with the context. For historians of France, "representation" (qua the act of one person speaking for another) has become the organizing principle for explaining the massive political upheaval separating the eighteenth century from the nineteenth. For philosophers since Kant, "representation" (qua the image one has of a thing in one's mind) has marked that troublesome point of mediation between the world as it is and how it appears to us. Within science studies, "representation" (qua the practice of producing inscriptions) has emerged as the key site in examining how scientific knowledge travels.

Arago and Biot, who spent their lives in and out of both science and politics, found ample opportunities to involve themselves in all three forms of representation. "Alas!," wrote Frédéric Chopin in 1848, "Everything is going wrong in this world. Think only that Arago with the eagle on his breast now represents France!!!"[10] Arago had just won the election for the head of the Executive Committee of the Second Republic. The fact that he would now somehow represent France, rather than the recently deposed King, was testament to the changing nature of the concept. This shift in the very meaning of the word "representation" has been at the heart of the "critical turn" examining France's revolutionary past through the creation of its symbols.[11] Moreover, many works in French history have fruitfully shown the shift in representation as a broadly cultural movement not confined to politics alone. Studies of theater, sign language, and print images have revealed the multiple venues in

[10] Frederic Chopin to Grzymala; London, October 17–18, 1848, in Frederick Niecks, *Frederick Chopin as a Man and Musician* (London: Novello, Ewer & Co., 1890), vol. 2, 304.

[11] Antoine de Baecque, "The Allegorical Image of France, 1750–1800: A Political Crisis in Representation," *Representations* 47 (Summer 1994), 111–143; Roger Chartier, "The New Press The World as Representation," Arthur Goldhammer and others, trans. in *Histories: French Construction of the Past*, Jacques Revel and Lynn Hunt, eds. (New York: The New Press, 1995); Lynn Hunt, *Politics, Culture, and Class in the French Revolution* (Berkeley: University of California Press, 1984); Maurice Agulhon, *Marianne au combat : L'imagerie et la symbolique républicaine de 1789 à 1880* (Paris: Flammarion, 1992); Keith Baker, "Representation," in *The French Revolution and the Creation of Modern Political Culture*, Keith Baker, ed. (Oxford: Pergamon Press, 1987).

which the French public rethought the way in which one thing stood in for another.[12] Paul Friedland, to take one example, has shown that theatrical and political representation underwent a "parallel evolution" in the last decades of the eighteenth century. On the stage, actors were no longer required to be their characters, but to present them in a manner the audience found believable. In the political realm, the *corpus mysticum*, which was thought to be all of France, was replaced by the elected assembly who only spoke "for" their constituencies in an abstract sense. In both instances, Friedland claims, a representation of embodiment gave way to one of *vraisemblance*.

But to draw all forms of representation together under the heading "one thing standing for another" leaves out a lot of the texture of what Arago and Biot were doing, both when they were depicting the world with their instruments and serving their terms as elected officials. "Representation," I claim, did not merely serve as a unifying metaphor linking disparate activities. Nor do I want to think of science and politics as parallel structures. These were not so much two similar things sitting side by side, as they were two different things that constantly found themselves overlapping.

So what, then, does it mean for one thing to stand in for another? Here, the discipline of science studies has shown us that there is nothing simple about that link. Symbols were things that needed to be made, and the people making them did so using materials and techniques. Inseparable from this act of making were questions about what the image actually was, how one invested the image with reliability, and who had the authority to make claims about what it was. Imaging technologies have long been a favorite subject of science studies, which provides us a methodological framework for how to analyze the physical activity of making visual images.[13] Representation has been opened up as a historical question, and it becomes interesting to inquire into the specific conditions that allow certain images to stand in for some piece of nature.[14]

The mechanism of the interaction between the political and the epistemic, this book asserts, can be located at the level of materials and manipulations

---

[12] Paul Friedland, *Political Actors: Representative Bodies and Theatricality in the Age of the French Revolution* (Ithaca: Cornell University Press, 2002); Sophia Rosenfeld, *A Revolution in Language: The Problem of Signs in Late Eighteenth-Century France* (Stanford: Stanford University Press, 2001); Sheryl Kroen, *Politics and Theater: The Crisis of Legitimacy in Restoration France, 1815–1830* (Berkeley: University of California Press, 2000); Joan B. Landes, *Visualizing the Nation: Gender, Representation, and Revolution in Eighteenth-Century France* (Ithaca: Cornell University Press, 2001).

[13] Michael Lynch and Steve Woolgar, eds., *Representation in Scientific Practice*, (Cambridge, MA: MIT Press, 1990). Bruno Latour and Peter Weibel, eds., *ICONOCLASH: Beyond the Image Wars in Science, Religion and Art* (Cambridge, MA: MIT Press, 2002).

[14] Peter Galison, *Image and Logic: A Material Culture of Microphysics* (Chicago: Chicago University Press, 1997); Lorraine Daston, "Objectivity versus Truth," in *Wissenschaft als kulturelle Praxis, 1750–1900*, Hans Erich Bödeker, ed. (Göttingen: Vandenhoeck & Ruprecht, 1999): 17–32; Timothy Lenoir, ed., *Inscribing Science: Scientific Texts and the Materiality of Communication* (Stanford: Stanford University Press, 1998); Robert Brain and M. Norton Wise, "Muscles and Engines: Indicator Diagrams and Helmholtz's Graphical Methods," in *Universalgenie Helmholtz: Rückblick nach 100 Jahren Hrsg. Von Lorenz Krüger* (Berlin: Akademie Verlag, 1994).

of the representing practices. People had to make the representations which circulated in French society. And the question of how they were made had implications for who could make them. The technical debates about the polarimeter centered on the question of whether or not one needed Newtonian algorithms to know what colors one was seeing. To ask this was also to ask whether the person using the instrument did or did not need a mastery of the algorithms in question. This question was not an abstract one for the two men. As they brought the polarimeter with them into other spaces, they deployed it as a tool for forming particular kinds of communities. Transparency and obscurity thus become concrete mechanisms of inclusion and exclusion.

Science and politics after the revolution

As two young men seeking their fortune in Paris, Arago and Biot looked a lot like the ambitious young heroes of a Stendhal novel. Both men had, like Stendhal, issued from families of mid-level functionaries (Arago's father was the treasurer of the mint at Perpignan; Biot's was an official at the National Treasury). Both moved from provincial obscurity to the city of Paris, seeking glory through the newly founded École Polytechnique (Stendhal, by the way, had applied but had not made it in). In Paris, they fell under the wing of the same powerful mentor, Pierre Simon Laplace, who introduced them to the inner circle of scientific life. Their successes followed one after another quickly and gloriously. Within a few years, the two young men, still friends, found themselves members of France's most prestigious institutions.

Biot and Arago forged their scientific careers in a world that had been vastly reshaped by the Revolution. The scientific community they joined was itself brand new, having taken firm root in the wake of revolutionary reorganizations.[15] The revolution opened the modern era in both politics and science, the two activities energized by a "wellsprings of vigor" that fed them both.[16] A social transformation had broken apart the old system of status and replaced it with the promise of mobility. This new focus on what one does, rather than what one is, impacted the scientific as well as the civic realm, producing a shift from the self-educated general savant to career physicist shaped by France's new, professional institutions. Arago and Biot were perfect examples of this new generation of professional physicists, founding their careers on talent rather than birth.

Historians of science have well documented these early years, as the two navigated strikingly similar career paths.[17] Indeed, the two of them formed

[15] Nicole Dhombres and Jean G. Dhombres, *Naissance D'un Pouvoir: Sciences et Savant en France, 1793–1824* (Paris: Editions Payot, 1989).

[16] Charles Coulston Gillispie, *Science and Polity in France: The Revolutionary and Napoleonic Years* (Princeton: Princeton University Press, 2004), 2.

[17] Robert Fox, "The Rise and Fall of Laplacian Physics," *Historical Studies in the Physical Sciences* 4 (1974): 89–136; Maurice Crosland, *The Society of Arcueil: A View of French Science*

something of a pair, as Laplace's promising students benefiting from his patronage and connections. Their similar trajectories were all the more striking for their doctrinal differences: Biot was well known to be among the most conservative members of the scientific community, and Arago the most radical. Their very companionship seemed a testament to the irrelevance of national politics to the young scientists scrambling to establish their careers. And historians have generally treated the scientific community as itself largely politically neutral.[18] Science had many political *uses*, but these were independent of the particular side using them. Pierre Simon Laplace and Joseph Louis Lagrange functioned primarily as figureheads during the Empire, and scientists were afforded even less access to power during the reactionary Restoration. The practitioners of science preferred to cast their work as morally and politically neutral. Arago and, particularly, Biot even made gestures in this direction, cultivating the belief that their participation in science made them objective and above the fray.

But were science and politics really so distinct? The answer to that question requires, somewhat paradoxically, a step away from a social-structural analysis emphasizing institutional restructuring and changes in career path. I propose we look instead at the cultural act of attaching meaning to the world. Both Arago and Biot established their reputations on their ability to depict the world with optical instruments. Like so many newcomers making their way in the upheaval of the revolution, they were creating their world through a form of symbolic action. It was the disagreement over how this was done, how they assigned meaning to their optical products, which marked their greatest divergence. This book traces the path of this divergence, starting with the issue of transparency in their work in optics, and mapping through its effects more broadly on cultural, social, and political questions. The goal is not to reduce their scientific beliefs to their politics or vice versa. There was no single substratum that serves as an explanatory key for their actions. There was instead a continual layering, where the scientific and political moved back and forth in dialogue over the possibility of representation and cultural production.

This book takes the onion, with its many layers, as its structural model. The chapters have a rough chronology, but they also try to capture a thematic layering in Arago and Biot's own work. The chapters build upon one another, with each stratum extending further into French culture. Chapters 1 and 2 form the innermost layer. These chapters lay out the technical details of their disagreement over the nature of light. The pressing issue of the day (rather than the particle vs. wave debate, which only became important retrospectively) was whether all colors could be broken down by a complex analytical technique provided in Newton's *Opticks*. This question leapt to the fore when Arago and

at the Time of Napoleon I (Cambridge, MA: Harvard University Press, 1967); Eugene Frankel, "Career-Making in Post-Revolutionary France: The Case of Jean-Baptiste Biot," *British Journal for the History of Science* 11 (1978): 36–48.

[18] Dhombres and Dhombres, *Naissance D'un Pouvoir*.

Biot each developed their own version of polarimeter, a device that detected polarized light and produced a brightly colored display. Biot insisted that the colors could only be described using Newton's techniques, whereas Arago resisted the process of analyzing them. Behind their mathematical and experimental arguments was a philosophical difference: Biot's insistence on analysis implied that the eye was incapable of distinguishing colors on its own, that without the regulatory aid of his techniques, observers did not know what they were observing. Arago, on the other hand, maintained a greater confidence in the eye's ability to tell the difference between different combinations of colors, and in the observer's ability to see the world as it was.

Chapter 3 adds another layer to their debates over light by focusing on their work on astronomy. The only access each man had to their astronomical subjects was, of course, through the medium of light. But what was the nature of this light, and what did it reveal? The controversial question of celestial influence hinged upon the possibility that the light of the heavens brought with it more than what was visible to the eye. Arago and Biot fell on different sides of the question. When Arago became director of the Paris Observatory, he spent a great deal of effort debunking what he called superstitious myths of celestial influence. Using his optical instruments, he studied the radiations of comets, the moon, and other heavenly bodies, ultimately claiming they could have no appreciable effect on Earth. Biot, on the other hand, immersed himself in the study of ancient Chinese and Egyptian astronomy. He was particularly engrossed with the zodiacs of both cultures, which he represented as a moment of profound original knowledge hinting at a relation between heavens and earth.

Chapter 4 treats the question of light and living bodies. Arago once again fashioned himself as a debunker, questioning the claims of the rising spiritualist movement that there exist previously unknown forms of radiation that act on living organisms. Biot, meanwhile, strove to make exactly that point: the world could be divided into active (living) and inactive (nonliving) matter. One could only distinguish the two by the effect they had on the plane of polarization of light. Their work participated both in the broader polemic of materialism and their long-standing disagreement over the nature of light and what it revealed.

Chapter 5 involves the use of light in the making of images. Both Arago and Biot were central figures in the early history of photography in France. They, with Alexander von Humboldt, were on the first scientific committee to report on Louis Daguerre's work. Here too, however, the two men ended up fighting. Each took up the side of one of the rival processes: Arago promoted Daguerre's silver plates and Biot supported William Henry Fox Talbot's paper photographs. Wrapped up in this disagreement about technique was a disagreement about the relationship of the photograph to the world. For Arago, the daguerreotype's ability to represent was guaranteed by the striking resemblance between the image and the world. For Biot, his photographic papers did not represent at all, but registered an invisible world of chemical radiation inaccessible to the human eye. Further tied to the disagreement about technique was the question of inclusivity and exclusivity. For Arago, the daguerreotype's transparent visibility meant that anyone who looked at one knew what they were seeing. For Biot, the only way to understand a photograph was to know the complex chemical processes involved in its creation. Not just anyone, he claimed, could have access to the secrets it revealed.

Chapter 6 treats the two men's involvement in colonial trade and slavery. Biot developed an optical device that became a crucial part of the colonial sugar industry, used to determine sugar quality. Arago, for his part, was a tireless critic of the plantation system. As the Minister of the Colonies in 1848, he signed into law the definitive abolition of slavery in the French Colonies. In both cases, issues of social organization centered around the question, "is man free?." Here we have arrived at one of the most pressing political issues of the day. It would make little sense to say that their thoughts on these matters depended in any direct way on their optics. Yet a thread runs clearly through the body of their lives' work: the question of human freedom depends on whether or not one is free from the influence of unseen forces. Whether one is free from unseen forces depends on whether or not light can be purely apprehended by its visible component, or whether there was more to it than one could see. This question, in turn, went straight to the heart of their original work: what, precisely, did light reveal about the world? This is not to imply that this thread constitutes an airtight logical chain or a unidirectional causal arrow, but it does point to the way the most technical of scientific questions and the most general of political questions get enfolded into one another.

1
A Revolution in Representation

Arago and Biot spent December 1806 together on a chilly mountaintop in Spain. Their makeshift camp at the Desierto de las Palmas was a small hut whose roof kept blowing off in the wind. As Biot wrote back to the Bureau des Longitudes that had sent them:

> We lead here a very retired life, having only for neighbors on our mountain a few Chartreux monks who never speak, and a few eagles who come from time to time to glide above our quarters. These quarters, completely wild as they are, appear superb to us, as we have there the certitude of being able to fulfill the task entrusted to us.[1]

The task in question was measuring the meridian line used to determine the length of the meter. Young and full of ambition, Arago and Biot had hit upon the expedition to Spain as their path to scientific glory. After suggesting it to Laplace, and receiving his support, they set off as the two-man triangulation team.

Their camp established, Biot left Arago at Desierto de las Palmas to travel to neighboring peaks. They lit fires at night, and peered through the darkness for each other's signals. They then measured the angles between peaks, adding these new triangulation points to the meridian chain. Arago spent six months alone this way, with little to do during the day. He found, to his delight, that the monks took neither their vows of silence nor most of the other precepts of Christianity too seriously. The monks also had a habit of nipping into the consecrated wine, even offering him a glass in the middle of a service. Arago was, he reported, a bit shocked at first. But not wishing to cause offence, he drank his glass of wine. Biot, however, would not swallow the blasphemy so easily. On his return, he argued so heatedly with the monks that one went off to get his pistol. It was only with great difficulty, Arago reported, that he persuaded the monk to spare Biot's life.[2]

The episode was a small one. Arago probably only mentioned it, years later in his memoirs, because of the delicious irony of saving the life of the man who would go on to serve as a constant thorn in his side. But the detail of

---

[1] Biot, "Au desierto de la Palmas, le 7décembre 1806," MS 1054, Bibliothèque de l'Observatoire.

[2] François Arago, *Histoire de Ma Jeunesse* (Paris: Christian Bourgois Éditeur, 1985), 76.

consecrated wine signals a demarcation. What, after all, did each man see in the glass that Arago eventually accepted and Biot nearly lost his life over? Wine or blood of Christ? Friendly drink or sacrilege? Transubstantiation had become, with the French Revolution, a political issue. The ancien regime had followed what Roger Chartier called the Eucharist model of representation, where the physical body of the king fused together with the mystical body politic in a process essentially sacramental. Louis XIV's "l'État, c'est moi" carried the same function as Christ's "This is my body."[3] Yet competing models of representation emerged in the eighteenth century to challenge the traditional logic.[4] The king's body, transformed by divine fiat into the body of France, was replaced by a popular will transparently representing the nation.[5] A revolution in representation had transformed the relation between signifier and signified. At the heart of the Revolution was a shift in the meaning of representation, a change in how one thing stood in for another.[6]

This chapter deals with two areas of science rewritten by the Revolution: metrology and color theory. The revolutionary Commission of Weights and Measures proposed the universal meter as a new standard of measurement independent of reference to the king's body. At another revolutionary institution, the École Centrale des Travaux Publics (later renamed the École Polytechnique), the school's faculty (many of whom were also on the Commission of Weights and Measures) substantially revised previous methods of treating colored bodies. The question of representation was central to both of these scientific projects, as revolutionary savants attempted to pin down the relation between the world and depiction, and struggled with the question of how to come to agreement in the absence of a single external authority.

As Arago and Biot established their careers in the aftermath of the revolution, they inherited this fractured legacy of representation. They became involved, as young students, in both the meter project and the new color theories of the École Polytechnique. The first case marked their only episode of cooperation, as they jointly made measurements establishing the meter. The question of color, on the other hand, provoked a split that endured throughout their careers. Even as they were shining light beams back and forth to one another in Spain, they had different ideas of what light was, and what kind of representation it allowed.

[3] Roger Chartier, *The Cultural Origins of the French Revolution*, Lydia G. Chochrane, trans. (Durham: Duke University Press, 1991), 129.

[4] Keith Baker, "Representation," in *The French Revolution and the Creation of Modern Political Culture*, Keith Baker, ed. (Oxford: Pergamon Press, 1987).

[5] Sara E. Melzer and Kathryn Norberg, eds., *From the Royal to the Republican Body: Incorporating the Political in Seventeenth- and Eighteenth-Century France* (Berkeley: University of California Press, 1998).

[6] Keith Michael Baker, "Representation Redefined," in *Inventing the French Revolution: Essays on French Political Culture in the Eighteenth Century* (Cambridge: Cambridge University Press, 1990), 224–251; Paul Friedland, *Political Actors: Representative Bodies and Theatricality in the Age of the French Revolution* (Ithaca: Cornell University Press, 2002).

Background

Biot began performing missions for the Bureau of Longitudes in 1803. He had been pulled to Paris from the provinces by Laplace, the head of the Bureau and the most powerful figure of French scientific society. Biot had caught the eye of Laplace as a young professor of mathematics and astronomy at the École Centrale de l'Oise in Beauvais, where he had been teaching from 1797. In a stroke of audacity, he had written to Laplace in 1799, as the *Mécanique céleste* was being printed, offering to double check the calculations of the page proofs.[7] The two men entered a lively correspondence. When a position at the College de France opened in 1800, Laplace pushed hard for Biot's appointment and the young man was soon on his way to the capital.[8]

Biot's first mission was to investigate claims of flaming stones falling from the sky in northern France. Tales of meteors had been common in the eighteenth century, but the scientific community usually dismissed the claim of their celestial origin (and the accompanying assumption of their function as signs from above). Biot's trip marked the first effort supported by the Academy of Sciences to investigate the phenomenon in depth. In 1803, he reported to the village of l'Aigle, where villagers had seen a luminous globe streak across the sky, then explode into numerous pieces which fell to the ground. He interviewed the witnesses, visited the site, and brought back samples of the meteors to analyze the chemical composition. His first contribution to science sided in favor of "one of the most surprising phenomena that mankind has ever observed": stones had indeed rained down from the heavens as the villagers had reported.[9]

The next year, in 1804, he was, with Gay-Lussac, the first to ride a balloon in the name of scientific experimentation.[10] The two aeronauts mounted to 4000 meters, running tests on the atmospheric composition and the intensity of the earth's magnetic field. They billed their ascent as the conquering of a new territory, the atmosphere, and capitalized on the substantial excitement generated by the new and dangerous sport of ballooning.[11] Danger was a constant element of scientific expeditions. That same year, 1804, the director of the Observatory, Pierre Méchain, died in the service of science. He contracted

---

[7] Biot, *Une anecdote relative à M. Laplace: lu à l'Académie française, dans sa séance particulière du 5 février 1850* (Paris: Impr. nationale, 1850).

[8] Eugene Frankel, "Career-Making in Post-Revolutionary France: The Case of Jean-Baptiste Biot," *BJHS*, 11 (1978): 36–48.

[9] Biot, "Relation d'un voyage fait dans le Département de l'Orne pour constater la réalité d'un météore observé à l'Aigle le 6 floréal an 11," reproduced in Jean-Paul Poirier, ed. *Jean-Baptiste Biot & la météorite de l'Aigle* (Paris: l'Académie des Sciences, 2003).

[10] Biot, *Relation d'un voyage aérostatique, fait par mm. Gay-Lussac et Biot; lue à la classe des sciences mathématiques et physiques de l'Institut national, le 9 fructidor an 12* (Paris, 1804). "Extrait du Moniteur"; printed from the same setting as Gazette nationale ou Le Moniteur universel, Paris, 1804, vol. 30, [1499]–1500.

[11] The Montgolfier brothers had invented ballooning only twenty years earlier. Richard Gillespie, "Ballooning in France and Britain, 1783–1786," *Isis*, 75 (1984): 249–268; Charles Coulston Gillispie, *The Montgolfier Brothers and the Invention of Aviation* (Princeton: Princeton University Press, 1983).

malaria while on an expedition to measure the meridian and perished before he could get back to France.[12]

Méchain's death set in motion the events that brought Biot and Arago together. The duties of Observatory secretary had been filled by Méchain's son, who withdrew after his father's death. With the position vacant, Siméon-Denis Poisson mentioned the name of Arago, whom he had met at the École Polytechnique, as a replacement. Arago was at first resistant. He had entered the École Polytechnique with dreams of a military career, and still had hopes of fulfilling it. A visit from Laplace, however, persuaded him. Arago signed on as the secretary of the Observatory, while still maintaining his name on the role of the École Polytechnique in case he wished to return to the military.[13]

Arago's first assignment at the Observatory was a collaboration with Biot. They undertook, at the suggestion of Laplace, a project determining the refraction of light through a gas.[14] Laplace had derived an equation, in Book 10 of the *Mécanique céleste*, for the refraction of light through the atmosphere as a function of altitude and temperature. He had based his work, however, on the assumption that the refractive force was proportional to the density of the air, and it was this assumption he wanted his assistant astronomers to test. They took to their project with zeal. Although there was a brief flare-up at the end when Biot tried to leave Arago's name off the publication, troubles were soon smoothed over and their names appeared side by side on the memoir read before the Academy of Sciences on March 24, 1806.

It was while working together that they hatched the plan of redoing Méchain's measurements. The meter project loomed large in the imaginations of both men. Arago remembered meeting Méchain in Perpignan on the voyage south that would eventually claim his life.[15] Just the year before (1803), Biot had written a history of the revolution that condemned the upheaval, but praised the role of scientists in establishing the meter. The meter was, for Biot, one of the few redeeming features of the French Revolution.

The meter was a complicated symbol that pointed to the difficulties of how a brand new nation was going to rule itself. The old *toise* of the ancien regime had, famously, been based on the size of the king's foot.[16] With the toppling of the king, however, came the question of how to establish a new standard of measure. The question was that of the Revolution writ large: how do people organize agreement among themselves in the absence of arbitrary authority? In an effort to ground the legitimacy of its decision on nature itself,

---

[12] Ken Alder, *The Measure of All Things: The Seven-Year Odyssey and Hidden Error that Transformed the World* (New York: The Free Press, 2002), 284.

[13] Arago, *Histoire* , 58.

[14] Biot and Arago, *Mémoire sur les affinités des corps pour la lumière* (Paris: Baudouin, 1806).

[15] Arago, *Histoire*, 144.

[16] Specifically, the cubit measure of the toise was six times the length of the King's foot. In practice, the standard was an iron bar publicly displayed at Châtelet. Charles Couson Gillispie, *Science and Polity in France: The Revolutionary and Napoleonic Years* (Princeton: Princeton University Press, 2004), 227.

the National Convention assigned a commission drawn from the Academy of Sciences the task of inventing a rational system of measurement. The commission defined the meter as one ten-millionth part of a quarter of the circumference of the earth (called the meridian). On the one hand, the commission went out of its way to establish the meter as a natural length, drawn from the earth, rather than from something arbitrary, such as the size of the king's foot.[17] But the end result was a platinum bar that would itself be invested with an absolute authority. There were thus two definitions of the meter: one taken from nature and one in the state archives. This dual nature points to one of the tricky parts of the revolution: the notion of using nature as the standard of agreement implied that nature was transparent and directly accessible. But, of course, the length of one ten-millionth of the meridian was not immediately available to everyone; hence, the need for the platinum bar. The gap between these two things, nature and its representation, might not have been a big deal. Indeed, it was part of the necessary fiction of the Revolution that it was not. But bad luck struck. For after the platinum bar had already been cast and ensconced in the National Archives, the astronomers discovered a mistake in the measurement of the meridian.

The mistake was the fault of Méchain, the same astronomer Arago had met as a boy. His task had been to head south from Paris, measuring the meridian as far as Barcelona. Meanwhile, the other half of the expedition, Delambre, went north, reaching as far as Dunkirk on the northern coast of France. They made their measurements through a process called triangulation, based on the premise that if the length of one side of a triangle is known, as well the angles between the sides, the length of the other sides can be calculated. Delambre and Méchain thus strung together a ladder of triangles, each sharing one side. After measuring a first baseline distance, they could determine the length of their entire ladder. Things went more or less smoothly for Méchain until he reached Barcelona, where he was to complete the triangulation with a reading on Mount Jouy, a fortress slightly outside the city. While he was there, war broke out between Spain and France. Expelled from the fortress, but not allowed to leave the city, Méchain tried to make the most of his time by observing from his hotel terrace. He thus wound up with two sets of latitude measurements for Barcelona: the terrace of the hotel where he was staying and the hill of Mount Jouy. But the data did not agree. He calculated a discrepancy between them, so large, that one or the other had to be wrong. He did not know which, and he had already mailed in the Mont Jouy observations.

Méchain kept quiet about the discrepancy as the results were compiled and the definitive length of the meter was announced.[18] As Delambre pressed him for his

---

[17] There is some debate about whether the commission truly believed in the possibility of naturalistic measures or were simply playing politics. John Heilbron, "The Measure of Enlightenment," in *The Quantifying Spirit in the Eighteenth Century*, Tore Frängsmyr, John Heilbron, and Robin Rider, eds. (Berkeley: University of California Press, 1993), 207–242; Gillispie, *Science and Polity in France*, 238.

[18] Which remained largely unknown until revealed in Alder, *Measure* .

notes, however, it became clear he could not perpetuate the secret forever. Hoping to undo the consequences of his error without rendering it public, Méchain proposed in 1801 to embark on another expedition, this time extending the arc of the meridian past Barcelona, as far south as the Balearic Islands, Ibiza and Majorca. This way the measurement for the arc would bypass Barcelona, and no one would notice that the value he had given for it was slightly off. To the Bureau of Longitudes, he gave the argument that the new arc would extend beyond the 45th parallel, and thus make the extrapolation from partial arc to meridian more reliable and less dependent on the earth's eccentricity.[19] Yet it was this effort at redemption that cost him his life, as he contracted malaria as soon as reaching Barcelona.

The failed expedition hung like a talisman before Arago and Biot, and they often spoke of it while working together at the Observatory.[20] They proposed to Laplace that they take up the mission. Laplace nominated them to the Bureau des Longitudes, who then accepted on May 2, 1806.[21] A few months later, armed with Méchain's old notebooks, Arago and Biot set off for Spain. They followed the route taken by Méchain. After picking up his abandoned instruments in Valencia, they established their observation camp in Desierto de las Palmas. The mountain offered an unobscured view south to the other mountains in the chain and west across the Mediterranean to the Balearic Islands. For the next six months, Arago remained on his mountaintop while Biot moved down the coast, ascending to the highest peaks he could find, and exchanging signals with Arago. They lit fires, amplified by an arrangement of mirrors, and peered into the darkness to find the fire of the other. They then used a Borda repeating circle to measure the angles between triangulation points. The greatest challenge came when Biot made it to Ibiza, and lit a fire on the mountaintop of Campvey for Arago to see back on the mainland. The stretch across the open ocean was considerably longer than any of the other distances, and Méchain had indicated that he thought the task was impossible. Arago and Biot employed eight mirrors instead of the usual two or three, and still it took six months before Biot finally spied Arago's signal (Arago never did see Biot's).[22]

The two spent a second winter, less chilly than the first, along the Spanish coast. In the spring, they headed to the islands to complete the triangulations between Ibiza, Majorca, and Formenta (see Figure 1.1). Biot then returned to Paris with the data, which he presented to the Bureau des Longitudes on March 3, 1808.[23] Arago stayed behind in Majorca to make some additional measurements to determine the curvature of the Earth.

Arago had planned to return in the next few months, but world history intervened. Relations between France and Spain were deteriorating quickly. The

[19] Ibid., 266.

[20] Arago, *Histoire*, 61.

[21] Procès-verbaux de Bureau des Longitudes, Observatoire de Paris, MS 1022. p. 70. Gillispie, *Science and Polity in France*, 477.

[22] *Rapport fait au Bureau des Longitudes, en sa Séance du Mercredi 1ere Juillet 1807; par les Commissaires chargés de la continuation des travaux pour la prolongation de la Méridienne de France aux Iles Baleares*, Archives Nationales, f17, 3713, M9.

[23] Procès-verbaux de Bureau des Longitudes, Observatoire de Paris, MS 1022. p. 75.

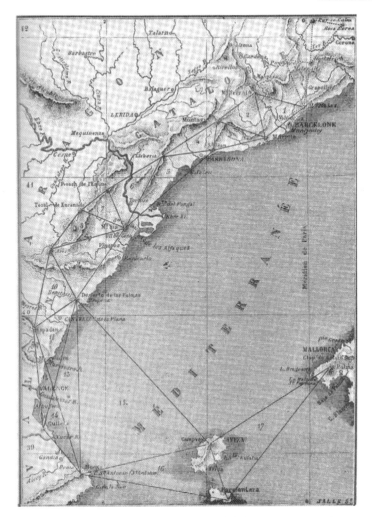

Fig. 1.1 The triangulation performed by Biot and Arago extending the meridian south along the Spanish coast and Balearic Islands. (From Biot and Arago, *Recueil d'observations géodésiques, astronomiques et physiques, exécutées par ordre du Bureau des Longitudes de France, en Espagne, en France, en Angleterre et en Écosse, pour déterminer la variation de la pesanteur et des degrés terrestres sur le prolonge-ment du Méridien de Paris, faisant suite au troisième volume de la Base du Système métrique* (Paris: Courcier, 1821).)

rumor began to circulate that Arago, camped on a high hilltop above the harbor, was a French spy secretly signaling the army waiting to attack.[24] When an officer of Napoleon did arrive, on May 27, 1808, an angry mob went looking for Arago.

---

[24] Arago, *Histoire,* 80. Although Arago's own memoirs are virtually the only account of his adventures, many of the details have been independently verified. Paul Murdin, "'Laborious

Fearing for his life, Arago fled for the nearby prison, with the Majorcans only a few steps behind him and succeeding in getting in a few blows.

He arranged to escape on a fishing boat to Algeria, where he picked up a fake passport claiming he was from Hungary. The Dey of Algiers sent a ship to France, with Arago on board, on August 13, 1808. They were in sight of Marseille when they were accosted by a Spanish corsaire. They were taken back to Spain, where Arago spent another several months in prison. By September, the Bureau still had no idea where he was. Forced to confront the possibility that he might be dead, they voted to suspend his salary.[25]

It would not be the Bureau of Longitudes, but the Dey of Algiers who came to Arago's rescue. Arago had written a letter telling him that the Spanish stopped the ship and killed one of his lions. The Dey demanded the release of his crew and they left on November 28, 1808. This time, however, nature got in the way. The mistral blew them off course and they landed back in Africa, in Bougie. Too impatient to wait three months for the next ship to Algiers, Arago disguised himself as a local and set out overland. He had joined in the Islamic prayer to avoid detection, guided, he recalled, by the farcical scene in Molière's *Bourgeois gentilhomme*.[26] When he arrived in Algiers, he learned that his protector the Dey had been decapitated. In February 1809, moreover, the new Dey declared war on France over an unpaid debt. Arago's name, along with those of other French nationals, was added to the list of "slaves of the Regency." Although, he admitted, his life as a slave was a rather comfortable one as the guest of the Swedish diplomat. It was not until June that the financial disagreement was worked out and Arago was permitted to leave, after six months in Algiers. He arrived in France on July 2, 1809.

Reunited in France, Biot and Arago continued their warm association. They became the closest friends in Paris of the famous adventurer Alexander von Humboldt, known for his extensive travels and wide-ranging studies of the Earth.[27] Humboldt had just returned from a long trip through the Americas and was newly installed with the Prussian diplomatic mission in Paris. He had befriended Biot soon after the latter's balloon ascent. Biot helped him perfect his writing in French, and provided the mathematical analysis for his reams of terrestrial magnetism measurements. Humboldt was equally struck by news of Arago's adventures in Spain. Upon hearing of the younger man's safe return, he dashed off a letter which was the first thing to greet Arago on French shores. Although Humboldt's friendship with Biot would fade,

and perilous adventures': François Arago's triumphant return to France," *Journal for Maritime Research* (2006).

[25] Procès-verbaux de Bureau des Longitudes, Observatoire de Paris, MS 1022. p. 77. The assumption of his death was further supported by a striking coincidence. When in prison, Arago had sold his watch to procure food. The watch eventually fell into the hands of a [French officer] passing through Perpignan. Arago's father recognized it, and assumed it could only mean that his son was dead. Arago, *Histoire*, 99.

[26] Arago, *Histoire*, 111.

[27] The literature on Humboldt is extensive. For an overview, see Nicolaas Rupke, *Alexander von Humboldt: A Metabiography* (Frankfurt: Peter Long Publishing, 2005).

his relationship with Arago grew into what one biographer called "the great passion of Humboldt's life."[28] The two men lived together from 1809 to 1811 (when Arago married), and maintained an intense correspondence through letters after Humboldt left Paris in 1827.

Arago and Biot soon found themselves colleagues at the Academy of Sciences. Biot had won his election in 1803. The next election took place on September 18, 1809, only weeks after Arago returned from his ordeal. He obtained 47 of the 52 votes, which should have made his election a clear victory except for the fact that Laplace, the most influential voice of the Academy, had thrown his weight behind Poisson, who was also up for election. Arago's supporters set to work trying to persuade Laplace to accept the nomination. At the head of the ranks were Biot and Humboldt, who was also the first to hear the news of Arago's eventual victory. He rushed to the Observatory to wake everyone up, then quickly dashed off a note to Biot to let him know of their victory, writing, "The joy was not small, and in order to share it with you I write these lines."[29]

Arago also joined Biot as one of the younger members of the Société d'Arcueil. This group started in 1807 as an informal gathering at the houses of Laplace and Berthollet, who were neighbors in the upscale Arcueil suburb of Paris.[30] Together, they served as patrons to a small group of the most promising young men of science. Biot was able to attend the first meeting, held in July 1807, because he had returned briefly to Paris from Spain to replace a circular scale that had broken in transit. He also resumed attendance when he returned permanently in 1808. As soon as Arago was back in Paris, he, too, began attending, and was there for the most active period of 1811–1813.

Arago and Biot thus launched their careers in tandem, publishing together, risking their lives together, and supporting one another's nominations. By 1811, however, the good will between them had evaporated. They moved from collaborating to attacking. Instead of supporting each other's nominations, they deliberately sabotaged attempts to advance. What happened? The standard answer was that Biot and Arago were engaged in a case of professional rivalry, where two similarly matched opponents competed for the limited rewards of their chosen profession.[31] Their rivalry reached its peak over the subject of polarization. Malus had discovered the phenomenon of polarization by reflection in 1809, but had died within three years of his discovery, leaving the field wide open. Arago and Biot, both well versed in optics and looking for a new topic, rushed in, each hoping to control the new domain.

But a certain amount gets lost in this telling. They were, after all, not merely fighting over priority. There was a substantive difference to their

[28] L. Kellner, *Alexander von Humboldt* (London: Oxford University Press, 1963), 81.

[29] Humboldt to Biot, Paris, October 20, 1809, Bibliothèque de l'Institute MS4896.

[30] Maurice Crosland, *The Society of Arcueil: A View of French at the Time of Napoleon I* (London: Heinemann, 1967).

[31] Eugene Frankel, "Corpuscular Optics and the Wave Theory of Light: The Science and Politics of a Revolution in Physics," *Social Studies of Science*, 6 (1976): 141–184.

arguments. They each had a different idea of how light worked, about how light revealed the world to the observer. To begin to tease apart the differences between them, we can look at the first solo works that they published in optics. For Biot, this was a translation of Ernst Fischer's textbook in mathematical physics, published in 1806. Most of the translation, from German to French, was done by his wife, Gabrielle. Biot himself provided an introduction and footnotes. In only one instance did he disagree with Fischer, however. This was on the subject of an optical phenomenon known as Newton's rings, which was the sequence of colored rings produced when two lenses were pressed against one another. Fischer had claimed the colors produced in these rings were simple, and Biot added a footnote to insist that they were, as Newton had claimed, complex.

Arago's first publication was an essay on the same subject, entitled "On the Colors of Thin Films," but unlike Biot, he came to bury Newton, not to praise him. His observations, Arago claimed, showed certain flaws in Newton's system. Like Fischer, he called into question Newton's explanation of the colored rings of thin films. When Arago and Biot moved into the study of polarization, their disagreement centered over whether or not Newton's analytical theory of color based on thin films could be applied to the colors of polarized light.

Color and the *Education des Artistes*

One might think that the colors of Newton's rings were an obscure topic, but it was not. By the time Arago and Biot were writing, the topic was the most important one in the field of optics. No optical subject was more condemned in Revolutionary France. Before becoming a radical Jacobin, Jean-Paul Marat had devoted his life to the study of light and vision. In the 1780s, he came out with a series of memoirs attacking Newton's theory of colors.[32] Johann Wolfgang von Goethe, watching from Germany, compared the tyranny of Newton to that of the French monarchy.[33] He spoke of the destruction of Newton's color theory as "razing this Bastille."[34]

But one does not have to look to the fringes of science to find a critique of Newton's colors. Indeed, within the very institutions that formed Arago and Biot, there was a division of opinion. Both men received remarkably similar educations, both formally, at the École Polytechnique, and informally, at the Société d'Arcueil. But even within this small group of like-minded spirits, all dedicated to the project of producing a rigorous mathematical physics, there were fundamental divisions over Newton's color theory. Laplace and

---

[32] Marat, *Mémoires académiques* (Paris: N.T. Méquignon, 1788); Jean Bernard, Jean-François Lemaire, and Jean-Pierre Poirer, *Marat, homme de science?* (Paris: Collection Les empêcheurs de penser en rond, 1993).

[33] Myles Jackson, "A Spectrum of Belief: Goethe's 'Republic' Versus Newtonian 'Despotism'," *Social Studies of Science,* 24 (1994): 673–701.

[34] Cited in Jackson, 677.

Berthollet, the two hosts of Arcueil, disagreed on the issue of color. With the École Polytechnique, as well, faculty members included both Newton's most loyal adherents and his toughest critics.

Throughout most of the eighteenth century, the question of why objects were colored the way they were was answered with reference to Newton's *Opticks*. Newton devoted a good portion of this work to a discussion of the generation of colors. In the *Opticks*, he had discussed the colors of natural bodies as a physical property related to the microscopic composition of their surfaces. Newton's own views on the matter were in fact rather nuanced, and in other places in the *Opticks*, he made the claim that colors were the sensations produced in the observer. This aspect of his thought, however, was not usually part of late eighteenth-century discussions of "Newton's system." The two textbooks published at the turn of the century that claimed to set out Newton's theory of colors (Haüy's *Traité élémentaire de physique* and Biot's *Traité de physique expérimentale et mathématique*), both made strong claims for the strictly physical nature of color.[35] The following discussion is taken from Biot.

Newton's theory of colors rested primarily on his work with the colored rings produced by pressing two lenses together. Usually called "Newton's rings" or the colors of thin films, they were the result of the small air gap existing between the lenses. As the thickness of the gap varied, the color changed. Newton explained this phenomenon by his theory of "aethereal vibrations" or fits. The ether within the gap would undergo successive periods of condensation and rarefaction, or fits. If the ether was condensed when the particle reached the second surface, it would be reflected back. If rarefied, it would be transmitted. For a single color, the result was alternating bands of light and dark. For white light, the various bands would overlap and mix together, and produce a repeating succession of colors.

Newton worked the theory out in full mathematical precision. By using a lens of known curvature to form the rings, he could use simple geometry to calculate the distance of the gap at any position.[36] He then observed the rings formed by isolated spectral colors, and for each color measured the diameters of the rings produced. He found that each color had a different "thickness" or diameter to its rings. For example, he determined that 14 blue rings could be produced in the same space as 9 red rings.[37] He also found that for each color, the succession of rings followed a mathematical consistency: if he squared the diameters they formed the arithmetical progression of odd numbers.

Newton had provided a table which gave the succession of colors produced by white light, and the corresponding thickness of the air gap. Biot reproduced the table, complete with Newton's values, in his *Traité* (see Figure 1.2).

[35] Réné Just Haüy, *Traité élémentaire de physique* (Paris: Imprimerie Delance et Lesueur, 1803); Biot, *Traité de physique expérimentale et mathématique* (Paris: Deterville, 1816).

[36] For a description of the techniques Newton used for this calculation, see Alan Shapiro, *Fits, Passions, and Paroxysms: Physics, Method, and Chemistry and Newton's Theories of Colored Bodies and Fits of Easy Reflection* (Cambridge: Cambridge University Press, 1993), 53.

[37] Ibid., 61.

	COULEURS RÉFLÉCHIES.	ÉPAISSEURS DES LAMES en millionièmes de pouce anglais,		
		d'air.	d'eau.	de verre.
1ᵉʳ ORDRE.	Très-noir	$\frac{1}{2}$	$\frac{3}{8}$	$\frac{10}{11}$
	Noir	1	$\frac{3}{4}$	$\frac{20}{11}$
	Commenc. du noir.	2	$1\frac{1}{2}$	$1\frac{4}{7}$
	Bleu	$2\frac{2}{5}$	$1\frac{4}{5}$	$1\frac{11}{10}$
	Blanc........	$5\frac{1}{4}$	$3\frac{7}{8}$	$3\frac{4}{5}$
	Jaune	$7\frac{1}{9}$	$5\frac{1}{3}$	$4\frac{1}{7}$
	Orangé	8	6	$5\frac{1}{6}$
	Rouge	9	$6\frac{3}{4}$	$5\frac{4}{7}$
2ᵉ ORDRE.	Violet	$11\frac{1}{6}$	$8\frac{3}{8}$	$7\frac{1}{7}$
	Indigo	$12\frac{1}{6}$	$9\frac{5}{8}$	$8\frac{1}{11}$
	Bleu............	14	$10\frac{1}{2}$	9
	Vert	$15\frac{1}{8}$	$11\frac{1}{3}$	$9\frac{1}{7}$
	Jaune	$16\frac{2}{7}$	$12\frac{1}{5}$	$10\frac{1}{7}$
	Orangé	$17\frac{2}{9}$	13	$11\frac{1}{9}$
	Rouge éclatant....	$18\frac{1}{3}$	$13\frac{3}{4}$	$11\frac{1}{6}$
	Ecarlate.........	$19\frac{1}{3}$	$14\frac{1}{2}$	$12\frac{2}{3}$
3ᵉ ORDRE.	Pourpre	21	$15\frac{3}{4}$	$13\frac{11}{20}$
	Indigo...........	$22\frac{1}{10}$	$16\frac{4}{7}$	$14\frac{1}{5}$
	Bleu............	$23\frac{2}{5}$	$17\frac{11}{20}$	$15\frac{1}{10}$
	Vert	$25\frac{1}{5}$	$18\frac{9}{10}$	$16\frac{1}{4}$
	Jaune	$27\frac{1}{7}$	$20\frac{1}{3}$	$17\frac{1}{2}$
	Rouge..........	29	$21\frac{1}{4}$	$18\frac{5}{7}$
	Rouge bleuâtre....	32	24	$20\frac{2}{3}$
4ᵉ ORDRE.	Vert bleuâtre.....	34	$25\frac{1}{2}$	22
	Vert............	$35\frac{2}{7}$	$26\frac{1}{2}$	$22\frac{3}{4}$
	Vert jaunâtre.....	36	27	$23\frac{3}{7}$
	Rouge..........	$40\frac{1}{3}$	$30\frac{1}{4}$	26
5ᵉ ORDRE.	Bleu verdâtre.....	46	$34\frac{1}{2}$	$29\frac{2}{3}$
	Rouge.	$52\frac{1}{2}$	$39\frac{3}{8}$	34
6ᵉ ORDRE.	Bleu verdâtre.....	$58\frac{3}{4}$	44	38
	Rouge..........	65	$48\frac{3}{4}$	42
7ᵉ ORDRE.	Bleu verdâtre.	71	$53\frac{1}{4}$	$45\frac{4}{7}$
	Blanc rougeâtre...	77	$57\frac{1}{4}$	$49\frac{1}{3}$

Fig. 1.2 Table of the colors of thin films (or Newton's rings). Biot took this table, numbers and all, directly from Newton's *Opticks*. It allowed him to assign a number to each color that appeared in the sequence of Newton's rings. (From Biot, *Traité de physique expérimentale et mathématique* (Paris: Deterville, 1816), 4: 77.)

Biot also repeated Newton's warning that these colors were not simple spectral colors. Nor did the designation "blue" mean that the spectral blue light was most intense at that point. The color was, of course, the composite of all the various rings of the spectral colors which, because of different diameters, overlapped with one another. Using the mathematical relations he had established, Newton provided an algorithm for going back and figuring out the "ingredients" that made up these compound colors.[38]

---

[38] Shapiro goes into the details of this technique, called the nomograph (which Biot mentions in the *Traité*), Shapiro, *Fits, Passions*, 94.

The natural colors of opaque bodies, claimed Newton, had an analogical relationship to the colors of thin films. An object derived its color from the particular rays that it reflected. The laws that determined which rays would be reflected were exactly the same as those that governed colored rings. Newton further proposed that the physical process was itself the same. That is, light particles incident upon the surface of an object would be reflected according to their fit state of condensation or rarefaction, and the particular fits that were reflected, depend on the corpuscular structure of the body involved. To describe the color of an object, then, one needed to pass through the algorithms that Newton had drawn from his work on thin films.

By the time Biot published his textbook, however, the Newtonian theory of color had been under attack from a variety of different disciplines. A wave of criticism of "Newton's theory" appeared in France in the 1790s. One must be precise, however, about what this term meant. Few people objected to his work with the decomposition of white light or even the phenomenological account of Newton's rings. What did come under fire, however, was the claim that the colors of natural bodies could be treated like the colors of Newton's rings, and thus explained in entirely analytic terms. Claude Berthollet launched this critique in his 1791 textbook, *Les éléments de l'art de la Teinture*.[39] This work was not the most notable document of 1791, as that year also saw the appearance of the first constitution of France. But it marked a revolution nonetheless. For the previous 100 years, the study of optics had been dominated by a single authoritative tone: Newton's *Opticks*. Although claiming to feign no hypotheses on the nature of light, Newton had argued hard that colors were compound and could be analyzed by analogy to the colors of thin films. It was this assumption that Berthollet called into question. One must not, Berthollet claimed, confuse the fugitive colors of oil slicks and peacock feathers with the constant colors conserved through changes of density and thickness. While the former followed the rules of Newton's rings, the latter did not. Their color properties, according to Berthollet, were easiest explained by the bodies' particular affinity for different rays of light. Berthollet also rejected the claim, implicit in Newton's analogical theory, of the compound nature of color. He held that most of the natural colors of bodies must be simple, because they had a simple cause and it was impossible to separate them.

Berthollet was also one of the organizers of the École Centrale des Travaux Publics, the school founded in 1794 to quickly educate a group of technically competent engineers for the revolutionary cause.[40] The republic needed engineers fast for the war effort, and most of the students had scanty mathematical background, if any. The response of the founders was to propose an entirely new kind of technical education, bypassing the complex rules of analysis and focusing on visual instruction intended to convey the world in as

---

[39] Claude-Louis Berthollet, *Les éléments de l'art de la teinture* (Paris: Firmin Didot, 1791).

[40] The other members of the committee appointed by the National Convention were Gaspard Monge, Prieur de la Côte d'Or, J.A. Chaptal, A.F. Fourcroy, L.B. Guyton de Morveau, N.S. Vauquelin, and J. H. Hassenfratz. Michelle Sadoun-Goupil, *Le chimiste Claude-Louis Berthollet (1748–1822): Sa vie son œuvre* (Paris: J. Vrin, 1977), 36.

straightforward a manner as possible. Its defenders called for an "education of the artisan and worker" which relied heavily on descriptive geometry and drawing. The key to this visual education was that, by depicting the world directly, one alleviated the need for difficult analytical manipulations. Analysis became the enemy of direct representation.

The "Cours révolutionnaire" was the heady title given to the first series of lessons taught in 1794.[41] Its revolutionary nature was, as the school itself was designed to be, at once both scientific and political. The democratization of technical education would be achieved, it was claimed, through a new science of representation. The school was "the first and only official instruction consecrated to the arts of drawing."[42] And drawing is what students spent most of their time doing. Indeed, they spent almost as much time in their drawing course as in all of their other courses put together. The curriculum was organized around the ten-day week of the revolutionary calendar. Every day except for quintidi and décadi, the students spent their mornings in 1-hour lectures on descriptive geometry, chemistry, and analysis. On quintidi, they attended 1-hour lectures on drawing and physics. But in addition to this, they also spent 3 hours every evening, between 5 pm and 8 pm, at a drawing practicum.[43]

Color was one of the key elements of this revolution. When the *Cours révolutionnaire* opened in 1794, several of the instructors made the critique of Newtonian color a central part of their courses. Color was treated, within the curriculum, not as an abstract theory but as part of the engineer's practical toolkit for visual communication. Color was both part of the project to replace equations with pictures, and subject itself to the epistemic shift from head to hand. The central target was Newton's analytical approach. As the physics professor, Hassenfratz, stated in 1795, the study of color, "stagnant from the instant that Newton ended his research, has expanded considerably at the École Polytechnique in the past six months, by the numerous experiments which have been done there."[44] In courses as diverse as drawing, descriptive geometry and physics, a single complaint emerged: the analytical approach to color was insufficient. Each took aim at the Newtonian theory of color as excessively abstract and authoritarian.

---

[41] The school's name was changed to the École Polytechnique on September 1, 1795. For a history of the school's origins, see Bruno Belhoste, *L'École Polytechnique et ses élèves de la Révolution au Second Empire* (Paris: Belin, 2003). Ken Alder, *Engineering the Revolution: Arms and Enlightenment in France, 1763–1815* (Princeton: Princeton University Press, 1997); Janis Langins, *La République avait besoin de savants: les débuts de l'École polytechnique— l'École centrale des travaux publics et les cours révolutionnaires de l'an III* (Paris: Belin, 1987); Bruno Belhoste, Amy Dahan Dalmedico, Dominique Pestre, and Antoine Picon, eds. *La France des X: Deux siècles d'histoire* (Paris : Economica, 1995); Ambroise Fourcy, *Histoire de l'École Polytechnique*, introduction by Jean Dhombres. (Paris: Belin, 1987); Terry Shinn, *Savoir, politique et pouvoir social, L'École polytechnique: 1794–1914* (Paris: Presses de la Fondation nationale des sciences politiques, 1980); Bruno Belhoste, Amy Dahan Dalmedico, and Antoine Picon, eds., *La formation polytechnicienne: 1794–1994* (Paris: Dunod, 1994).

[42] Edouard Pommier. *L'art de la liberté: Doctrines et débats de la Révolution française* (Paris: Gallimard, 1991), 283.

[43] Langins, *La République*, 29.

[44] Jean Henri Hassenfratz, "Physique," *JEP*, 2 (1794): 373–408.

The drawing course, where students spent the bulk of their time, sought to incorporate color into a practice of immediate expression and move away from Newtonian abstraction. Of the nine lectures the drawing professor François-Marie Neveu gave during the *Cours révolutionnaire*, one was devoted almost entirely to the subject of color. Its central argument was that Newton's color system was excessively abstract and of little use in depicting the world. In particular, Neveu argued against Newton's classification of colors into a seven-part scale. Four of these seven colors, he pointed out, were simply mixtures of the other three. And indigo, which Newton claimed to be a fundamental color, was clearly just a shade of blue. Just because the two shades appeared at opposite ends of the spectrum did not mean that they were fundamentally distinct. Neveu was hardly the first to mount this objection; many of the Academic painters he disparaged similarly rejected Newton's seven-color taxonomy. But Neveu further linked his objections to his general critique against the corruption of the ancien regime.

Newton had gone wrong, Neveu claimed, by founding the system on an analogy with the seven-note scale in music. As Newton pointed out, the proportion of the lengths of the intervals between the colors corresponded closely to those of the musical tones. The highest tones were the sharpest, as the highest colors were the brightest. But, claimed Neveu, this analogy was flawed because of a fundamental difference between seeing and hearing. Placing an obstacle between the object and the observer completely annihilated vision, while doing little to one's ability to hear. An image, unlike sound, can only be experienced by direct action. Physicists who followed Newton, Neveu accused, were too quick to take the visible as the image of the invisible, the effect as the image of its cause. They wanted to study nature to find what it did not reveal, and thus made the error of separating the material and intellectual, and considering nature apart from its existence.[45] This abstraction was, for Neveu, a form of perversity. The "direct action" of visual experience required a color system unmediated by mathematical systems.

The same critique was echoed in the course of Descriptive Geometry, taught by the mathematician Gaspard Monge. Monge had been the guiding hand behind the École, and his course, descriptive geometry, lay at the heart of the program for the *"éducation des artistes."* It was through the practice of engineering drawing that artisans, unacquainted with the rigors of analysis, could visually communicate their knowledge. And it was within the context of engineering drawings that Monge introduced the subject that had formed the basis of much of his research for the past ten years: color.

In 1789, Monge published an article attacking the claim that the color of an object depended solely on the "absolute nature of its rays."[46] He chose not to pronounce on the question of whether the difference between colors lay in the

---

[45] François–Marie Neveu, "Dessin," *JEP*, 2 (1794): 130.

[46] Gaspard Monge, "Mémoire sur quelques phénomènes de la vision," *Annales de chimie et de physique* 3 (1789): 147.

nature of the light particles themselves, or in the different speeds at which they moved. Rather, he wanted to question the claim, assumed by both sides of the debate that the sensation of color lay in physical properties alone. He provided one example, "within reach of most observers," concerning the shadow cast by a candle on white paper.[47] Many had already noticed that, in the first light of the early morning, the shadow appeared blue. The typical explanation was that the shadow was being illuminated by light from the atmosphere, which was blue. But, Monge pointed out, when the candle was extinguished, the entire paper would be illuminated by the atmospheric light. And yet it appeared to the eye not blue, but white. After a series of similar observations, he ventured

> One would be led to conclude...that in the judgment that we make on the colors of objects, there enters, so to speak, something moral, & that we are not uniquely determined by the absolute nature of the light rays that the body reflects.[48]

Color could not be a strictly physical property, as the same species of ray seemed capable of exciting different perceptions of color. For Monge, the moral element of color was a necessary part of the practice of engineering drawing. The word "moral" here had a particularly eighteenth-century connotation, meaning something irreducibly human or pyschological. In the first lecture given at the *École Centrale*, he outlined the importance of this human element.[49] Descriptive geometry was the science of representing three-dimensional objects on two-dimensional pieces of paper. One way of accomplishing this was to provide two drawings of the object in different planes. Monge, however, preferred the use of shadows to provide a sense of depth. The correct coloring of these shadows, he continued, was the product of observation and experience. It depended not only on the properties of light, but on "causes we can regard as purely moral."[50] It was the neglected personal aspect of color perception, Monge said, that he hoped to develop further in the *Cours révolutionnaire*.

It was the physics professor Hassenfratz who gave the most thorough treatment of color and the most radical overhaul of Newtonian theory.[51] His first critique of Newtonian color was given as a lecture at the *Cours révolutionnaire* in 1794, and was then printed in the *Journal de l'École Polytechnique*. He began with the same phenomenon as Monge, that is, the blue shadows of sunrise. But then continued on that there was nothing particular about sunrise. Rather, every shadow that one observed was colored in some way, even those of high noon that looked perfectly black. Hassenfratz investigated the shadows

---

[47] Ibid., 135.

[48] Ibid.

[49] Monge, "Stéréotomie," *JEP* 1 (1794): 1–11.

[50] Ibid., 9.

[51] The term "physique" was at this time in transition between the eighteenth-century meaning of the general study of nature and the nineteenth-century meaning of a specific discipline. Alan Shapiro makes the point that most of Hassenfratz' research was in the area of chemistry, and he sided primarily with chemists on the subject of chemical affinities. Shapiro, *Fits, Passions*, 309.

produced by various light sources, and found that they took on every color of the prism. Like Monge, he concluded that these colors could not be solely the property of the physical particles.

Hassenfratz also noticed that not only were the shadows colored, but also they seemed to come in complementary pairs.[52] He observed the situation when there were multiple light sources in play, competing with the light of the sun or atmosphere. In this case, there were anywhere from two to six different colors visible in the shadows. If there were two colors, Hassenfratz claimed, it was always the case that these two colors were complements of one another. If there were three colors, one of them was always the complement of the two others. And so on for any number of shadows. One of these colors would always depend on the color of whatever object was providing the reflected light. For example, if the room was dominated by light reflecting off of slate surfaces, the shadows would be bluish. If the light was reflected off of plants or trees, the shadows would look green. All of this made sense, and Hassenfratz admitted that it could be easily predicted by current understandings of color and shadows. What was less predictable, however, was the presence of the second complementary color, which existed even in the absence of any body capable of reflecting that color.

Hassenfratz was careful to define the term "complementary color," because this was one of the first instances of its use in the French language.[53] He pointed out that the notion of complementary colors did not in any way oppose Newton's claim that white light was made up of an infinite number of homogeneous colors. This part of Newton's work, found in book one of the *Opticks*, was entirely compatible with his observations. It was the second part of Newton's color theory, found in Book 2 of the *Opticks*, with which Hassenfratz disagreed. This part had to do with the causes of color generation, and particularly the analogy with Newton's rings.

Hassenfratz published the full run of his critique "On the Coloring of Bodies" in the 1808 edition of the *Annales*. "Optics," he lamented, "has remained stationary since the publication of Newton's experiments on light."[54] The authority of the great man, he claimed, prevented serious research on even the most problematic aspects of the theory. In particular, Hassenfratz proposed to examine the central analogy posited between the natural color of objects and those created in the air gaps of Newton's rings.

Hassenfratz proposed to test Newton's theory by looking at the spectra of light that had passed through colored glass. According to Newton's theory, the spectrum of light that has been reflected from a colored body must be the same as the spectrum of a thin film of the same color. As the rules involved in putting these mixtures of light together were very precise, there were only certain

---

[52] Hassenfratz "Sur les ombres colorées," *JEP* 11 (1802): 276.

[53] Georges Roque, "Les couleurs complémentaires: Un nouveau paradigme," *Revue d'histoire des sciences et de leurs applications* 47 (1994): 405–433.

[54] Hassenfratz, "Mémoire sur la colorisation des corps," *Annales de chimie* 66 (1808): 152–67, 290–317; 67 (1808): 5–25, 113–51.

combinations that could occur. In the affinity theory, on the other hand, there were no laws governing the process of absorption, and any combination of colors could be expected to appear. Hassenfratz' task was thus to compare the spectra he obtained to those allowed by Newtonian theory. He found several discrepancies, and concluded that this theory alone could not by itself account for colors other than those produced in thin films.

Prieur de la Côte d'Or was at the time engaged in very similar researches. Although Prieur himself did not teach at the Polytechnique, he was one of the most important figures in its organization, and gave the opening remarks of the *Cours révolutionnaire*. He emphasized that the school distinguished itself by attaching more importance to "the work the student executes with his own hands" than to arid book learning.[55] This was, he assured, the best way of fixing knowledge in the spirit. Without manual labor, one's understanding of a thing would only be superficial.

After the revolution, Prieur undertook his major scientific work, "Considerations on color and their singular appearance."[56] He examined, as had Hassenfratz, the absorption spectra of opaque and transparent bodies and determined that the colors produced could not be generated by the laws that Newton had provided for thin films. Sometimes, he pointed out, absorption colors could not even be represented by the scale of thin film colors, as with certain violets of manganese oxide or indigo blues. Prieur thus joined his voice to the attack upon Newton, claiming that color was not a strictly physical property and could not be satisfactorily described by the analytical laws that Newton developed from his work with thin films.

The arguments made by Neveu, Monge, and Hassenfratz are clearly different in nature and have usually been placed within different disciplinary lineages. Neveu's statements against the seven-color scale can be seen as part of a long-standing tension between the artist and the physicist. Neveu's lecture would thus be one of the early hallmarks of the general shift in the early nineteenth century from seven to three colors that has been referred to as the "artists' revolt."[57] Monge's work with colored shadows is usually discussed as one of the first accounts acknowledging the subjective role of the observer in the process of vision, paving the way for Goethe's even more radical critique in *Die Farbenlehre* and the eventual establishment of physiological optics in the

---

[55] Prieur de la Côte d'Or, iii–viii "Avant Propos," *JEP*, 1 (1794): iii–viii.

[56] Prieur intended the work as a single text, which never appeared, although Prieur did present extracts of it before the Academy of Sciences on five occasions. Berthollet and Hassenfratz published those segments related to Newton's theory of colored bodies in the *Annales*. Claude Antoine Prieur-Duvernois, "Extrait d'un mémoire ayant pour titre: Considérations sur les couleurs, et sur plusieurs de leurs apparences singulières," *Annales de chimie*, 54 (1805): 5–27; "De la décomposition de la lumière, en ses élémens les plus simples; fragment d'un ouvrage sur la coloration," *Annales de chimie*, 61 (1807): 154–161; "Considérations sommaires sur les couleurs irisées des corps réduits en pellicules minces; suivies d'ue explication des couleurs de l'acier recuit, et de celles des plumes de paon. Fragment d'un ouvrage sur la coloration." *Annales de chimie* 61 (1807): 154–179.

[57] John Gage, *Color and Culture: Practice and Meaning from Antiquity to Abstraction* (Boston: Little Brown, 1993).

nineteenth century.[58] Hassenfratz, meanwhile, has been cited as part of the early history of using spectral decomposition as a means for chemical analysis.[59] Yet I would like to shift the focus from the vertical relations between the work and what came after it to the horizontal relations drawn between the projects in the particular location of revolutionary education. The curriculum of the École Centrale, designed by a commission that included Monge, Hassenfratz, Berthollet, Prieur, and Neveu, intended the subjects of descriptive geometry, chemistry, and drawing to serve as the foundation of a practical education that harnessed science to the service of the Republic. Knowledge, they claimed, must be put within grasp of all, and thus was better communicated through the visual language of engineering drawing than the abstraction of mathematical analysis. The work of Monge, Hassenfratz, and Neveu at the École was linked through a common goal of worker education, and color played a role in this project. To claim that color was immediately perceived and not computed through analysis was to render it accessible to the rudest of workmen.

Biot did not leave all of this out of his textbook because he was unaware of it. He had been one of the original students of the *Cours révolutionnaire*.[60] A little older and better prepared than the average student, Biot had been selected for a small group of *chefs de brigade* to help with instruction. As such, he participated in the *Cours'* highly touted project in which the advanced students conducted a work of original research. The project in question was none other than a study, guided by Monge, of the use of color shading to convey the three-dimensional character of engineering drawings.[61]

So what did Biot make of his revolutionary education? One can get a sense from Biot's 1803 "History of the sciences during the Revolution," one of the first attempts to write the revolutionary nature of the École Polytechnique out of history.[62] Science, claimed Biot in his history, had just reached a particular point of development before the Revolution. Throughout the seventeenth and eighteenth centuries, science had proceeded under the aegis of philosophy. Natural history, physiology, mineralogy, and other "sciences of observation" all rid themselves of prejudice and worked to spread Enlightenment.[63] By the end of the eighteenth century, however, sciences had "a new logic" and a new way of operating.[64] This logic paired together a new standard of experimental precision with rigorous reasoning methods. At the heart of these methods was what Biot called the path of "the moderns": analysis.[65] With this powerful tool, science wrested its independence from philosophy.

---

[58] For a discussion of the politics of Goethe's color theory, see Jackson, 673.

[59] Shapiro, *Fits, Passions*.

[60] Émile Picard, *La Vie et l'Oeuvre de Jean-Baptiste Biot* (Paris: Gauthier-Villars, 1931), 225.

[61] Citoyen Dupuis, "Mémoire sur la détermination géometrique des teintes dans les Dessins," *JEP*, 1 (1794): 167.

[62] Biot, *Essai sur l'histoire générale des sciences pendant la Révolution Française* (Paris: Duprat, Fuchs, 1803).

[63] Ibid., 6.

[64] Ibid., 22.

[65] Ibid., 24.

The Revolution, according to Biot, nearly halted this forward march of knowledge.[66] With both the universities and academies closed, the Republic's compulsion for egalitarianism seemed bent on dismantling centuries of accumulated knowledge.[67] The National Convention was dominated by what Biot called "the Revolutionary Party," who "only saw in the sciences a poison which irritated republics."[68] There was, however, a minority party, not blinded by politics, who argued in favor of organized science education. This minority group was able to triumph over "the ignorant and ferocious" by convincing them of science's utility to the war effort. The Republic needed to equip and train an army, and the National Convention created the École Polytechnique to do it. Science had saved itself by its ability to actually *do* things. It had, according to Biot, resisted the politicization of the Revolution by virtue of the independence bought by its technological applications.

Biot's rewriting of the history of the École Polytechnique was part of a battle over the fractured heritage of the Revolution. The founders had dreamed of a regenerated form of transparent communication. Yet while this program of worker education dominated the *Cours révolutionnaire* of 1794, it did not ultimately end up being the École Polytechnique's primary legacy. From 1795, analysis began to play a larger and larger role within the school's curriculum. Laplace, who had no role in the initial planning of the school, returned from hiding to put his own stamp upon it.[69] The drawing course was substantially curtailed. From 1795 to 1806, the percentage of time devoted to descriptive geometry dropped from 50% to 31%, while that of chemistry went from 25% to 9%.[70] Analysis, on the other hand, increased from 8% to 45%.[71] From 1795, the École Polytechnique carried within it a dual tradition in which geometry and analysis served as the foundations of two distinct mathematical and professional cultures.[72]

Biot found himself allied on the side of analysis. He found himself teaching the subject in 1799, when Laplace brought him back to the École Polytechnique as an *examinateur* of analysis. The introduction to his 1806 translation can be read as a declaration of what side he was on. The introduction took the form of a letter from Biot to Berthollet (one of the chief critics of Newtonian color), in

---

[66] Ibid., 34.

[67] Ibid., 36.

[68] Ibid., 50.

[69] Bruno Belhoste refers to this transition, which he pinpoints in 1795, as the shift from "l'école de Monge" to "l'école de Laplace." Belhoste, "Un enseignement à l'épreuve: l'École Polytechnique de 1794 au Second Empire," in *La formation polytechnicienne, 1794–1994*, Bruno Belhoste, Amy Dahan Dalmedico, and Antoine Picon, eds. (Paris: Dunod, 1994), 9–30. See also Gillispie, "Un enseignement hégémonique: les mathématiques," in *La formation polytechnicienne, 1794–1994* Bruno Belhoste, Amy Dahan Dalmedico, and Antoine Picon, eds. (Paris: Dunod, 1994), 31–44.

[70] Janis Langins, "The Decline of Chemistry at the Ecole Polytechnique (1794–1805)," *Ambix*, 28 (1981): 1–19.

[71] Ibid., 7; see also Jean Dhombres and Nicole Dhombres, *Naissance d'un nouveau pouvoir: sciences et savants en France* (Paris: Bibliothèque Historique Payot, 1989), 572.

[72] Lorraine Daston, "The Physicalist Tradition in Early Nineteenth Century French Geometry," *Studies in History and Philosophy of Science*, 17.3 (1986): 269–295.

which Biot presented the case that physics must be systematic and law-like, with a strong basis in analysis. These were precisely the properties that Newton's defenders claimed that their color theory possessed and Berthollet's affinities could not. Biot furthered his point by praising Haüy's recently published *Traité de physique*, which criticized Berthollet's color theory as unmathematical.[73]

Biot's most explicit defense of Newton's theory came in the footnotes. In Fischer's original text, the section on color was rather brief, and only mentioned Newton's rings as a means of decomposing white light. Here Biot added a note stating that Newton's explanation of the rings went much further than Fischer had indicated.[74] In particular, Fischer had implied that the colors produced were simple colors, which Biot pointed out was incorrect. Although the thin films decomposed the white light falling on it, they did not resolve it into simple rays. The colors produced were thus complex, and would show a compound spectrum if observed through a prism. Biot added another dissenting footnote to Fischer's statement that affinities played some role in the color of natural bodies.

> This manner of envisaging the phenomena indeed appears at first glance to be simpler; but when one goes further into it, one finds that it is infinitely less probable than that of Newton; This is what I hope to show elsewhere.

The "elsewhere" he referred to was most likely the *Traité de physique expérimental et mathématique*, his textbook on experimental and mathematical physics, which contained an extensive section on color.[75] The section defended the explanation of the colors of objects by analogy with Newton's rings, and criticized the recently proposed affinity theories. In it, Biot established himself as the last loyal proponent of the Newtonian system of colors.[76] Although intended as a textbook for general physics, its most pressing message was a defense of Newtonian color. And, to make his allegiances that much clearer, he returned to calling "colored rings," "Newton's rings," and adopted the seven-color scale, which included indigo as a fundamental color.

In Chapter 2, we shall see how the questions of the colors of thin films erupted into the well-known debate between Biot and Arago. Biot defended Newton's claim that the colors of natural bodies were analogous to those of thin films, and could only be treated analytically. Arago, meanwhile, took up the opposite side of the École Polytechnique's revolutionary legacy. Where Laplace had recruited Biot to teach the analysis course, Arago followed Monge, and began teaching geometry at the École Polytechnique in 1809.[77] Shortly after, he produced his first scientific paper, "On the Colors of Thin Films," which took up the critique of Newtonian color.

---

[73] Biot, "Introduction," in E.G. Fischer, *Physique Mécanique*, Jean-Baptiste Biot, trans. (Paris: J. Klosterman fils, 1806), xxi.

[74] E.G. Fischer, 445.

[75] Biot, *Traité de physique expérimental et mathématique* (Paris: Deterville, 1816).

[76] Shapiro, *Fits, Passions*.

[77] Arago, *Histoire*, 137.

2
Le Rouge et le Vert: The Colors of Opposition in Restoration France

The names François Arago and Jean-Baptiste Biot are best known in the history of science as the principal combatants in the optical revolution pitting the theory of light particles against that of light waves.[1] Yet looking at their work, one thing becomes clear very quickly: neither Arago nor Biot cared very much whether light was a particle or a wave. Arago, in fact, refused to pronounce on the matter at all, claiming that any statement on the subject was both premature and unnecessary. Biot, too, after a small number of early papers mentioning light corpuscles, argued emphatically and repeatedly that his optical work was independent of any assumptions about the nature of light. Casting the debate as one over particles and waves conveniently locates their work within a long trajectory stretching from Newton to Einstein. But it ultimately obscures what the actors themselves thought they were doing.

In *The Rise of the Wave Theory of Light*, Jed Buchwald began to move away from these "tools of explanation," or theories about the nature of light, to what he calls "tools of analysis," or the techniques used to solve optical problems.[2] The real revolution, he asserted, was the move from Arago and Biot's method of counting isolable rays to Augustin Fresnel's use of the wave front. Buchwald's insight does a lot of work in explaining the extended disagreement between Biot and Arago's protégé, Fresnel. It leaves unexplained, however, the dispute between Biot and Arago, who, despite a shared practice of ray-counting, emerged as the central antagonists in the highly public and acrimonious debates of the 1810s. This approach, too, is ultimately a form of

---

[1] See, for example, Xiang Chen, "Reconstruction of the Optical Revolution: Lakatos Vs. Laudan." *Proceedings of the Biennial Meeting of the Philosophy of Science Association* 1 (1988): 103–109; Eugene Frankel, "Corpuscular Optics and the Wave Theory of Light: The Science and Politics of a Revolution in Physics," *Social Studies of Science*, 6 (1976): 141–184; John Worrall, "Fresnel, Poisson and the White Spot: The Role of Successful Predictions in the Acceptance of Scientific Theories," in *Uses of Experiment: Studies in the Natural Sciences*, David Gooding, Trevor Pinch, and Simon Schaffer, eds. (Cambridge: Cambridge University Press, 1989), 135–157; Robert Silliman, "Fresnel and the Emergence of Physics as a Discipline," *Historical Studies in the Physical Sciences*, 4 (1974): 137–162.

[2] Jed Buchwald, *The Rise of the Wave Theory of Light: Optical Theory and Experiment in the Early Nineteenth Century* (Chicago: University of Chicago Press, 1989).

excavation, looking for hidden meanings below the surface of their words. Yet much can be gained by remaining on the level of their stated concerns, and making sense of their statements, not by finding some deeper explanatory key, but by placing them within the webs of meaning in which they operated.

So what were Arago and Biot fighting over? The paper that served as the center of the controversy, written by Fresnel in 1821, was entitled, "On the calculation of tints that polarization develops in crystal laminae."[3] What Arago and Biot disagreed over was, precisely, how to calculate the tints that polarization developed in crystal laminae. Arago had created, in 1811, an instrument, the polarimeter, that produced beams of complementary colored light by passing polarized light through crystal sheets. Biot soon devised a polarimeter of his own, and used it to come up with a set of intensity equations for the beams. He calculated the color coefficient of these equations by assuming that the colors were analogous to the colors of Newton's rings, and using the algorithms Newton had provided from these rings to calculate color. Arago rejected the analogy of the colors of the polarimeter with those of Newton's rings, and the implication that the colors of the polarimeter could only be determined using analytic techniques.

Their differences ultimately came down to an act of representation. Arago and Biot were arguing over the very concrete act of creating optical effects by using the polarimeter. This instrument rendered polarized light visible by separating it into two complementarily colored beams. For Biot, the polarimeter produced its colors in a manner analogous to the arrangement of Newton's rings, and he called upon Newton's analytical techniques to describe them. For Arago, this attempt to analyze, and the color theory it rested upon, was invalid. This question of practice also imposed certain limits on who was qualified to use the instrument. Could anybody look at these brightly colored images and describe what they saw, or did they need the algorithms of the *Opticks* at their disposal? What was needed to communicate information about what color the images of the polarimeter were? For Arago, it sufficed to look. The groups of people who gathered together in salons to watch his polariscopic displays all had unmediated access to the colors produced. They were all seeing the same thing, and could easily attach a color name to it and talk amongst themselves. For Biot, however, the claim that the images were not by themselves transparently interpretable ruined this possibility of unregulated communication. Because people did not know on their own what they were looking at, they were forced to have recourse to some external authority to coordinate their observations.

At the center of this intersection of representing practice and social organization is the issue of standardization. From virtually the moment Biot began working with a polarimeter, he conceived of using it as a means of standardizing colors. Each color could be assigned a number based on the thickness

---

[3] Augustin Fresnel, "Note sur le calcul des teintes que la polarisation développe dans les lames cristallisées," *Annales*, 17 (1821): 393–412.

of the mica in the polarimeter. Arago disagreed. The problem was precisely that Biot claimed that the polarimeter colors represented all the colors available in the world. He treated a simple spectral blue as indistinguishable from a blue that could be decomposed, for example, into a mixture of red, yellow, and violet light. For Arago, this amounted to the impossible claim that the eye did not know what it was seeing. This issue cut at once to the heart of the technical problem of determining the color coefficients, but was more than merely technical. As the complementary colors of the polarimeter circulated themselves within the worlds of fine arts and industrial dying, they brought with them questions of representation, color standardization, and how people were able to talk to one another.

Arago's new colors

Arago wasted little time in setting himself up as a critic of Newton. His first memoir to the Institute, presented on February 18, 1811, took three sentences to get to the point that current theories of light were "inexact or insufficient."[4] Usually, this has been read as meaning that the memoir posed problems for Newton's claim that light was corpuscular. Arago himself has encouraged this interpretation. In a note added to the memoir decades later, on the occasion of publishing his collected works in 1853, Arago recalled that he had indeed intended his work to decide between the particle and wave theories. He had, he claimed, even given a demonstration of the incompatibility of his results with the emission system. That part of the memoir, however, "was burned at the printers" and not mentioned again for forty years.[5] He made no effort to explain why the part of the memoir that survived the fire made unqualified, if infrequent, references to light molecules.[6] Read without the 1853 addition, Arago's attacks on "Newtonian theory" in 1811 look like something else entirely.

Arago's memoir, entitled "On the Colors of Thin Films," dealt entirely with Newton's explanation of the colors of thin films. Arago pointed out that Newton had devoted an entire book of the *Opticks* to these colors, and had used them as the basis for this theory on how objects appeared colored. The laws that determined which colors an object reflected, he claimed, were exactly the same as those that governed colored rings. Newton further proposed that the physical process was itself the same. The color of any given object would thus be a complex mixture of other colors, which could only be known by using the formulas Newton had provided for thin films. This necessary passage through analysis had recently come under attack by several physicists engaged in color research. Arago's memoir can be read as participating in this vigorous contemporary debate.

---

[4] François Arago, "Mémoire sur les couleurs des lames minces, lu le 18 février 1811," *Œuvres*, 10: 2.

[5] Ibid., 31.

[6] Ibid., 2.

Arago took particular aim at Newton's claim, closely related to the physical action of fits, that color was generated uniquely at the second reflecting surface of the thin films. According to Newton's theory, the crucial factor for deciding which colors were reflected was whether the ether was condensed or rarefied when it reached the surface. Arago, however, suggested that the surface might not be the only thing involved. He used, to support his claim, the brand new phenomenon of polarization by reflection. The year before, Etienne Malus had observed from the Observatory that the light reflecting obliquely off the windows of the Luxembourg Palace was polarized.[7] Previously, this phenomenon had only been observed in light that had passed through doubly refracting crystals. It could be detected by passing the light through a crystal of Icelandic spar, in which case a beam of polarized light would be split into two beams of different intensities that would vary with inverse proportionality to one another as the crystal turned. This property of the spar had made it commonly used at the Observatory as part of a common astronomical device, the *lunette prismatique*. This *lunette* contained within it a double refracting prism that produced two images of the object being measured.[8] Once having established the distance of separation of the two images by looking at an object of known length, Arago used it to measure such distances as the diameter of Jupiter or the width of Saturn's rings. Throughout the years 1810 and 1811, Arago took hundreds of micrometer measurements of celestial objects using a *lunette prismatique*.[9] He also began using the device as a double refracting analyzer in his polarization experiments, substituting it for the second piece of Icelandic spar in the experimental arrangement.[10]

When Arago looked at Newton's rings through the Icelandic spar, he was surprised to find that colored light behaved differently from white light. When Malus had compared the polarization of transmitted and reflected beams for ordinary light, he had found that the beams were polarized in opposite directions. But Arago found that for the colored rays generated in thin films, the two beams were polarized in the same direction. If, as Malus had proposed, the polarizing force occurred at reflection, there would be no reason to think that colored and white light would behave differently, as the colors were only generated themselves at the moment of reflection. Whereas the situation became less

[7] It had long been known that when light passed through a doubly refracting crystal, it emerged with a particular property called polarization. Malus extended the domain of polarization to include reflected light. See André Chappert, *Etienne Louis Malus (1775–1812) et la théorie corpusculaire de la lumière: traitement analytique de l'optique géométrique, polarisation de la lumière et tentative d'explication dynamique de la réflexion et de la réfraction* (Paris: J. Vrin, 1977); Jean Rosmorduc, *La polarisation rotatoire naturelle, de la structure de la lumière à celle des molécules* (Paris: Blanchard, 1983).

[8] For a description of the instrument, also called a *lunette de Rochon*, see D. Fauque, "Alexis-Marie Rochon (1741–1817), savant astronome et opticien," *Revue d'histoire des sciences*, 38 (1985): 23–35.

[9] Arago, "Mesures micrometriques faites à la lunette prismatique," manuscript notebook, Bibliothèque de l'Observatoire, Paris, MS E3(4).

[10] Arago, "Mémoire sur la polarization colorée, lu le 11 aout 1811," in Arago, *Œuvres*, 10: 36–74.

puzzling if one allowed that the colors were generated in the passage through the thin film, before the reflection occurs.

Arago next experimented with replacing the air with a thin sheet of mica.[11] The results fell even further afield of anything discussed in Newton's *Opticks*. As the mica was of uniform thickness, one would not expect to see colors generated by the method of fits. And yet Arago did see colors, very bright ones. "I saw at that instant," he described in a memoir of August 11, 1811, "that...for every position of the prism and the lamina, whatever the color of one of the rays, the second was always the complementary tint."[12] If he rotated the piece of mica, the images would turn successive colors in the order of Newton's rings, always remaining the complement of one another. The mica, claimed Arago, imparted a kind of modification on the light passing through it that he called "*polarization colorée*," or chromatic polarization.[13]

The term "complementary colors" was, we have seen, only recently coined by Hassenfratz, Arago's physics professor at the École Polytechnique.[14] Arago provided a definition that matched the use of the term by Hassenfratz: complementary colors were those that formed white when mixed together. Newton, Arago pointed out, had himself gotten complementarity wrong, by deducing from the phenomenon of colored rings the following list of "opposite colors":

red . . . . greenish blue
yellow . . . . violet
purple . . . . green[15]

Indeed, much of Hassenfratz's work on complementary colors took place within a critique of Newton's theory of color production by the method of fits. Hassenfratz, as we have seen in Chapter 1, rejected Newton's claim that the natural colors of bodies could be treated with the algorithms he gave for colored rings, and the associated claim that color was generated at the moment of reflection. Arago's 1811 work echoed this criticism when it claimed that color could not be simply a property of the microphysics of the reflecting surface, but must somehow be the result of "the particular action" of the thin film as the particle traversed it.

Arago soon incorporated the principles of chromatic polarization into the instrument of the polarimeter. This instrument was, at its most basic, a simple device to detect whether or not light was polarized. If light was passed through a mica sheet and a doubly refracting crystal and showed two images

---

[11] Arago did not discuss his reasons for doing this, but Buchwald points out that the mica would have served to isolate any given color of the ring (mica comes in uniform layers, and thus provides a uniformly thick air gap, which produces a uniformly tinted image). Buchwald, *The Rise of the Wave Theory*, 75.

[12] Arago, "Sur la polarization colorée," 10: 38.

[13] Rosmorduc, *La polarisation rotatoire naturelle*.

[14] He used it in 1794, the same year as the other candidate, Benjamin Thomson. Georges Roque, "Les couleurs complémentaires: Un nouveau paradigme," *Revue d'histoire des sciences et de leurs applications*, 47 (1994): 405–433.

[15] Arago, "Sur les couleurs complémentaires," in Arago, *Œuvres*, 10: 367.

of complementary colors, the light had been polarized. If the two images were of the same color, the light was unpolarized. Arago used this instrument in his daily astronomical activities. For example, on November 23, 1811, he reported that, while looking at the moon through the *lunette prismatique*, he noticed that one of the images was slightly brighter than the other. Suspecting polarization, he inserted a sheet of rock crystal (which he found produced the same results as mica). Sure enough, one image of the moon turned red and the other one green.[16] Moonlight, Arago could now report, was polarized.

Biot's defense of Newton

Within months, Jean-Baptiste Biot had begun his own work on the phenomenon of chromatic polarization. Both contemporaries and historians have regarded Biot's immediate and somewhat heavy-handed move into the field that Arago had opened as an inappropriate usurpation of a novel research subject. Yet there was a defensive element to Biot's actions as well: by 1811, he was already established as the principal defender of the theory of color formation that Arago was attacking.

Biot first entered the debate, as we have seen, with his 1806 translation of Ernst Fischer's textbook in mathematical physics. Although there he only gestured his disagreement with Fischer's critique of Newton, he promised to further elaborate in his own work. He was at the time compiling his own textbook on experimental and mathematical physics, *Traité de physique expérimental et mathématique*, which contained an extensive section on color.[17] Yet before Biot had completed the textbook, Arago's announcement of his new colors introduced a new threat to the integrity of the system. Biot's response was to try to bring the novel phenomena in line with the old. Within months, he produced his own statement on the subject, a work of several hundred pages that dominated its volume of the Laplace-controlled *Mémoires de l'Académie*.[18] The work meticulously described a long series of experiments all aimed at establishing one thing: that the colors Arago had discovered by using his polarimeter were precisely analogous to those of Newton's rings.

Biot's procedure, which he repeated many times, was to take a single crystal of calcspar, mica, or rock crystal (all doubly refracting) and slice it into laminae of varying thicknesses. He placed them in the polarimeter arrangement (between polarizing mirror and analyzer), and observed the colors produced. Observing the colors, however, involved more than just looking at them. The light forming the image was, he pointed out, some complex mixture of colors.

---

[16] Arago, "23 Nov 1811 (unattached pages)," manuscript notebook "Observations et notes," Bibliothèque de l'Observatoire, Paris, MS E3(13).

[17] Biot, *Traité de physique expérimental et mathématique* (Paris: Deterville, 1816).

[18] Jean-Baptiste Biot, "Sur des nouveaux rapports qui existent entre la réflexion et la polarisation de la lumière par les corps cristallisé," *Mémoires de l'Académie des Sciences (1 juin 1812)*. See also Buchwald, *The Rise of the Wave Theory*, 409–417.

The first step was to make this mixture equivalent to some simple spectral color using what he called "reduction by Newton's scale." This method was first presented by Newton in his *Opticks*. One arranged the colors in a circle, and assigned each of the colors involved a "weight" based on its intensity. One then added these together to find the center of gravity, which would lie in the circle at a point demarcating its simple color equivalent.

The next step, for Biot, was to turn this color into a number, and for this he turned to Newton's tables. These tables, first presented in the *Opticks*, listed in one column the thickness of the gap between two lenses, and in the other column the tint of the reflected light (see Figure 1.2 in Chapter 1). After recording these colors on a table, he went back and measured the thickness of the crystals, which he placed in the column next to the colors (see Figure 2.1). Before being able to compare these numbers to those of Newton's table, however, he first had to "reduce the measures." For example, he found that a crystal 33.8 thousandths of an inch thick produced a blue color of second order. Newton's table assigned this same blue the number 9. Biot, thus, went through and multiplied all of his thickness measurements by the factor 9/33.8. This number then became the "calculated color" that could be looked up on Newton's table and compared to the "observed color" (see Figure 2.2).

N.os des lames.	Leur couleur observée par réflexion dans l'azim. de 45°.	Leur mesure au sphéromètre.	Leur mesure réduite à l'échelle de Newton.	Nomb. qui appart. aux coul. les plus vois. dans la table.
0	blanc légèrement jaunâtre.	14,2	3,77	3,5 blanc du 1er ordr.
3	blanc bleuâtre.	11,6	3,09	*idem.*
2	jaune du 1er ordre.	17,8	4,74	4,6 jaune du 1er ordr.

Fig. 2.1 Table for calibrating the "calculated colors" of the polarimeter. The second column gives the color of the extraordinary ray. The third column gives the measured thickness of the analyzing crystal Biot used. The fifth column gave, for each color, the equivalent thickness of the thin film of Newton's rings (the numbers were taken directly from the third column of Newton's table, reproduced above as Figure 1.2). For dozens of measurements, Biot divided the number in column 5 by the number in column 3. He came up with an average of roughly 0.24. He then multiplied the number in column 3 by 0.24 to come up with the number in column 4, which was the thickness "reduced by Newton's scale." (From J.B. Biot, *Traité de physique*, 4: 358.)

Angle dièdre formé par le plan d'incid. avec le plan de polarisation primitif A.	Angle compris sur la surface de la lame entre son axe et la trace du plan d'incidence i.	Couleurs et limites d'intensité du faisceau ordinaire observées F_0.	Couleurs et limites d'intensité du faisceau extraordinaire F_e,	
			calculées.	observées.
0	0° 0'	blanc, maxim.	noir	noir
	22 30	lilas	jaune verdâtre	jaune verdâtre
	45	bleu très-beau, minimum.	jaune brillant, maximum.	jaune brillant, maximum.
	67 30	vert blanchât.	rouge orangé	rouge orangé
	90	blanc, maxim.	noir	noir
	90 + 22 30	vert blanchâtre	rouge orangé	rouge orangé
	90 + 45	bleu très-beau, minimum.	jaune brillant, maximum.	jaune brillant, maximum.
	90 + 67 30	lilas	jaune verdâtre	jaune verdâtre
	90 + 90	blanc, maxim.	noir	noir.

Au-delà de 180°, les teintes redeviennent les mêmes qu'à partir de 0°. C'est pour-quoi, dans les expériences suivantes, on n'a pas dépassé $i = 90°$.

45°	0° 0'	rouge violacé, minimum.	vert jaunâtre, maximum.	vert jaunâtre, maximum.
	22 30	blanc lilas	jaune verdâtre	jaune verdâtre
	45	blanc, maxim.	noir	noir
	67 30	vert bleuâtre	rouge orangé	rouge orangé
	90	vert très-beau, minimum.	rouge vif, max.	rouge vif, max.
90°	0° 0'	blanc, maxim.	noir	noir
	22 30	lilas	jaune verdâtre	jaune verdâtre
	45	bleu très-beau, minimum.	jaune brillant, maximum.	jaune brillant, maximum.
	67 30	vert blanchâtre	rouge orangé	rouge orangé
	90	blanc, maxim.	noir	noir.

Fig. 2.2 Table comparing the calculated and observed colors. Biot made thousands of observations at various thicknesses and angles comparing the color he observed (column 5) to the color that he calculated using Newton's table and the calibration factor determined in Figure 2.1 (column 4). The close match between the two colors in nearly all cases led Biot to claim that the colors of the polarimeter were analogous to the colors of Newton's rings, and could be analyzed in the same way. (From J.B. Biot, *Traité de physique*, 4: 383.)

He found the calculated color to match his observations in virtually all cases. He concluded from this that color and thickness "followed exactly the same relation as those Newton had observed" with the rings produced from ordinary reflection in thin films.[19] These results, he claimed, suggested "several new and very deep analogies between the still unknown causes which produce ordinary reflection in light, and those which polarize it in crystalline bodies."[20]

He provided a set of equations that gave the intensities of the ordinary and extraordinary rays as they emerged from the doubly refracting crystal.

$$F_o = O \cos^2 \alpha + E \cos^2 (\alpha - 2i)$$
$$F_e = E \sin^2 \alpha + E \sin^2 (\alpha - 2i)$$

where O and E were coefficients determining the ray's color for the ordinary and extraordinary rays, respectively. Biot had deduced these formulas using what Buchwald has characterized as the techniques of ray counting.[21] What could not be deduced, however, were the actual values for the coefficients O and E, or the colors of the light. Biot obtained these coefficients by reading them directly off of Newton's tables, after modifying by the two empirically established multiplicative factors. The numbers corresponded to single spectral colors, which were treated as equivalent to the actual complex mixtures of colors as the latter could be reduced to the former through the algorithm of Newton's color circle.

Arago resisted the inclusion of his phenomenon into the Newtonian system. He disagreed with Biot's central claim that the thicknesses of the crystal laminae could be reduced to those of the air layers in Newton's rings.[22] He procured for himself a large number of rock crystal lenses of all shapes, and set about investigating whether the colors produced depended only on the thickness of the crystal, or also on the intervals between the laminae. He was struck, in particular, by a lens that varied in thickness between 1 and 2 millimeters and yet could be made to produce a single color of red over its entire surface. "Or, in other words," Arago concluded, "the rule that Newton gave for ordinary rings should not be applied to those that I discovered in crystals, because the latter do not depend strictly on the material thicknesses of the bodies."[23] Several people, he said, had asked him for an instrument to use in public demonstrations. He had one made up that used sheets of rock crystal cut perpendicular to the ridges of the hexahedral prism that formed the principal accessory of the instrument. He thought he would see which thicknesses worked best, but he

[19] Biot, "Sur des nouveux rapports," 168.

[20] Ibid.

[21] Buchwald, *The Rise of the Wave Theory*, 95–98.

[22] Arago, "Mémoire sur plusieurs nouveaux phénomènes d'optique, lu le 14 décembre 1812 à la classe des sciences mathématiques et physiques de l'Institut Impérial de France," in Arago, *Œuvres*, 10: 85–98.

[23] Ibid., 92.

found that the thicknesses could vary between very extended limits without diminishing the intensity of the colors.[24]

Biot himself was quite concerned with the question of how the colors could vary as the angle of incidence changed. In his next memoir, another 371-page behemoth published in the *Mémoires*, he again turned to the analogy with Newton's rings to furnish an answer. After all, he pointed out, one observed incidence effects with ordinary colored rings. When light hit the air gap at an oblique angle, it had a longer distance to travel. The gap's effective thickness increased and produced a different color. Newton had discussed the effect in the *Opticks* and provided a formula to account for it. If e was the color produced by a particular gap at right angle, then one could find e', the color at oblique angles, using

$$e' = e/\cos u \quad \sin u = K \sin \theta'$$

where θ' is the angle of refraction and K is a constant.

In a notebook entitled "Calculations on Newton's Rings," Biot detailed his own criticisms of Arago's work. The central problem was that Arago was treating complex light made up of a mixture of colors as simple. He borrowed the red glass that Arago and Fresnel had used in their work on polarization and performed a few experiments of his own. Red glass was commonly used to get a homogeneous light that was easier to work with. Yet, Biot claimed, the rays transmitted by the glass could be shown through analysis to have quite different physical properties.[25]

The possibility of standardization

This debate was never far from the practical question of how to talk about color. Artists, dye-makers, and many others beyond physicists and chemists cared about how they could compare colors with one another. The issue of color standardization was a pressing concern. The polarimeter was immediately drafted into the effort, and became a central location of the debate between Arago and Biot.

In 1816, Biot modified the polarimeter to measure color, and renamed it the colorigrade. This device "realizes and fixes in an invariably constant and comparable manner, all the shades of color that natural bodies can offer."[26] The instrument that Biot presented to the Academy consisted of a black mirror placed in front of a tube, which could be turned with a screw such that its inclination reflected polarized light into the tube. He used a doubly refracting prism as the ocular lens. Between the mirror and the prism he placed a sheet of rock crystal cut perpendicular to the axis, which could be rotated (or Biot suggested mica as it flaked easily and thus obviated the skill required to cut

[24] Ibid., 95.
[25] Biot, Manuscript notebook, Bibliothèque de l'Observatoire, MSE3 55.
[26] Biot, "Construction d'un colorigrade," *Bulletin de la société philomatique* (1816): 145.

the crystal into thin sheets). As it was rotated, the extraordinary ray passed through the colors of Newton's rings. Biot once again relied on the table taken directly from the *Opticks* to assign, for each color, a number corresponding to the thickness of the air gap at that point. One could thus, according to Biot, "rigorously define" a color by the thickness of the gap of air that gave rise to that color on Newton's table. Biot also modified this arrangement to create a "cyanometer" which measured shades of blue from a pale blue-white to violet.[27] Biot made it clear that the instrument relied on exactly the point in contention. For Biot, the problem of reproducing all the colors of nature was the problem of reproducing the colors of Newton's rings, and this is precisely what he claimed the polarimeter did.

Arago was not convinced. Not only, he pointed out, did it require the analogy ("in my opinion rather hypothetical") that Newton made between the colors of his rings and natural bodies, it also required one to extend the analogy to the colors of the polarimeter. It seemed unlikely, he said, that Biot's instrument had all the generality it claimed. Some colors, Arago maintained, were simply not contained within the series of colors produced by the polarimeter. He gave the example of a certain piece of colored glass in his possession. The light it transmitted had a brilliant blue color, that, when decomposed with a prism, separated into distinct portions of two red, one greenish yellow, one-fourth green, one-fifth blue, one-sixth blue-violet, and one-seventh violet. It seemed impossible, to Arago, that this complex amalgam corresponded to any simple spectral color.[28] Certainly this color, Arago argued, would be unrepresentable by anything the colorigrade could produce.[29] To claim that the colorigrade reproduced it perfectly was to admit that the observer could not tell what he was seeing.[30]

For Biot this was a ridiculous objection.[31] After all, the colors produced by the polarimeter were themselves complex mixtures of different simple rays. There was, for Biot, no difficulty in moving easily between these complex mixtures and the tint they presented to the eye. One used the algorithms provided by another one of Newton's constructions, the color circle (see Figure 2.3).[32]

For example, a color of 1/4 blue, 1/4 red, and 1/2 yellow is represented by three spokes on the color wheel. These spokes are combined geometrically to yield a single line: the equivalent simple tint. Using the table Newton had drawn up for the colors of thin laminae, one could easily go back and analyze the different constituents of light that appeared simple to the eye. Arago had been quite right to point out that blue could be a mixture of reds and violets. His mistake had been to think that this blue would somehow look different

[27] Arago designed his own cyanometer, which he thought would be more useful for describing the color of the sky that used a single shade of blue, and varied the amount of white light mixed with it. Arago, "Construction d'un colorigrade—cyanométrie," in Arago, *Œuvres*, 7: 445.

[28] Ibid., 443.

[29] Ibid., 442.

[30] Ibid.

[31] Biot, *Traité*.

[32] Buchwald, *The Rise of the Wave Theory*, 409–417.

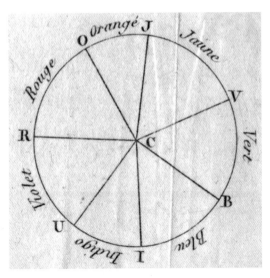

Fig. 2.3 A color wheel for giving the simple color equivalent of a complex mixture. The line G represents the geometrical combination of 1/4 blue, 1/4 red, and 1/2 yellow. (From J.B. Biot, *Traité de physique*, 4: pl. I.)

from the spectral blue, and could not be represented by the colorigrade. In fact, perception alone was incapable of distinguishing a complex and simple blue, only optical analysis could do that. When people spoke of seeing blue, they actually had no idea what they were seeing. That was why Newton's table was needed; it allowed people to talk about colors without having to know what they were.

Physicists were not the only ones interested in color standardization. Another figure grappling with the issue in post-revolutionary France was the painter Léonor Mérimée. As a *maitre de dessin* at the École Polytechnique, he had instructed both Arago and Biot in the art of drawing, and then taught beside Arago when the latter became professor of descriptive geometry. He had joined the group of Jacobin color theorists at the Polytechnique working under Neveu when the *Cours révolutionnaire* first opened in 1793. Prior to the Revolution, he had found little success in establishing his art career. He had competed several times for the Academy's *prix du Rome*, but after a string of second and third places he decided to travel to Italy on his own, where he associated primarily with other artists external to the Academy. In 1792, Mérimée received the order from Paris to replace the arms of Bourbons with those of the Republic on the French Consulate. The Roman authorities rewarded him with a brief stint in jail and an escort out of the country. He was feted in Paris, however, and installed in an apartment in the Louvre, where he also exhibited his work in the Revolutionary Salons.

The same year that Biot presented his colorigrade, Mérimée published his own system for communicating color. The venue was an 1815 treatise in

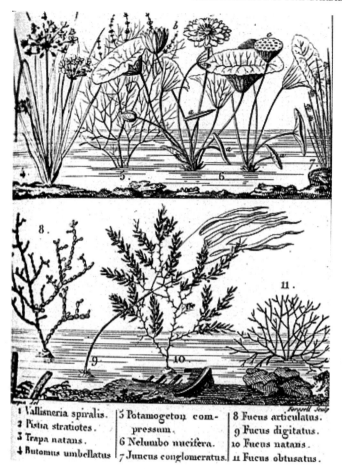

1 Vallisneria spiralis.	5 Potamogeton com-	8 Fucus articulatus.
2 Pistia stratiotes.	pressum.	9 Fucus digitatus.
3 Trapa natans.	6 Nelumbo nucifera.	10 Fucus natans.
4 Butomus umbellatus	7 Juncus conglomeratus.	11 Fucus obtusatus.

Fig. 2.4 Line drawings of plants. (From C.F. Brisseau-Mirbel, *Elémens de physiologie végétale et de botanique* (Paris, 1815).)

vegetable physiology. The treatise consisted primarily of line drawings of plants, such as the one shown in Figure 2.4.

It was of some importance, however, to include the color of a plant within its description, and Merimée was brought in to author a section discussing this problem.[33] His task was to develop a color scale that naturalists could use to standardize their descriptions. Yet Mérimée did precisely the opposite. He railed against efforts to standardize color descriptions through chromatometers, or collections of color samples typically used for standardization. These were unsatisfactory, Merimée pointed out, because the colors tended to fade or alter in fairly short order. He advocated, instead, that those who wanted to talk about color should make a color circle of their own. To this end, Merimée presented

[33] C.F. Brisseau-Mirbel, *Elémens de physiologie végétale et de botanique* (Paris: Magimel, 1815).

the rudiments of the color theory that held that all the colors of nature could be created from the three "generatives": red, yellow, and blue. These colors could be arranged in a circle, such that they formed the colors orange, green, and violet between them. Merimée also went further, providing circles with twelve divisions, including yellow-orange, red-orange, etc. (see Figure 2.5).

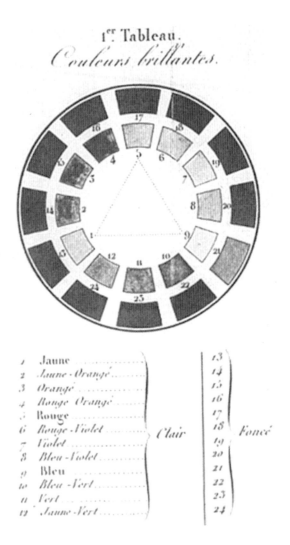

Fig. 2.5 Merimée's color wheel. The spokes of the triangle helped one mix complementary colors. (From C.F. Brisseau-Mirbel, *Elémens de physiologie végétale et de botanique* (Paris, 1815), 925.)

"It must be pointed out," noted Mérimée, "that by uniting any one of these twelve colors with that which is diametrically opposed to it, one forms the complement of the three generative colors...and if one mixes, in proper proportions, these opposing colors, one will have the same tint of an absolute gray, or else the circle is badly divided."[34] The best way to tell if one's own scale was accurate was to make sure the colors opposite one another were colorless when mixed together. Merimée's remark may seem a little banal, or at least in perfect keeping with the properties of the artist's well-known color wheel. But in fact, it represents one of the very earliest uses of the term "complément" within art literature.[35] And thus complementary colors became the means for subverting the project of standardization. After calibrating it with some easily available item, one could situate each color in question on its appropriate place in the scale. Mérimée thus proposed to multiply Newton's color wheel until each man had his own, and could describe a color within the framework of his own experiences.

The spaces of communication

Standardization was not a strictly technical problem for either artists or physicists. It was also a social one by which people coordinated themselves with one another so that they would be able to come to agreement. Arago and Biot disagreed over the use of the polarimeter for standardization because of a specific technical issue: whether the colors of the polarimeter could be treated analogically to those of Newton's rings. But this was also a disagreement over how easy it was to coordinate one's observations. For Arago, the colors were transparently available and each observer had only to look in order to see what color the images were. For Biot, the true color of the image could only be known by analyzing the complex color components, and coordination necessarily passed through Newton's tables of values.

These different views on communication were linked to different kinds of sociability. In particular, one can note a difference between a space outside any centralized authority, predicated on rational debate between equals and another that constantly refers back to some fixed standard to guarantee its meaning. The former requires the assumption of a certain transparency of communication, in which the coordination of perception is not a problem and everyone can understand one another without difficulty. The latter, on the other hand, takes into consideration the substantial work that often needs to be done to establish communication between them. These two different practices corresponded to different ways in which people arranged themselves in order to communicate. To understand the differences in how the polarimeter

---

[34] Ibid., 912.
[35] John Gage, *Color and Culture: Practice and Meaning from Antiquity to Abstraction* (Boston: Little Brown, 1993), 173.

functioned as an inscribing device, one must look at where it functioned. We turn, then, to the spaces that Arago and Biot traversed as they did their polarization work.

One place where Arago could be found throughout much of the 1820s and 1830s were the drawing rooms of the Paris salon. Complaining of "the insipid genre of life" at the Observatory, he would frequently spend his evenings with the artistic and literary society of Paris.[36] The papers hailed him as "one of the most celebrated talkers of the saloons. (*sic*)"[37] Stendhal, another salon regular, spoke warmly of Arago's contributions to evening conversation. In a travel essay written from Geneva, he expressed surprise that men would return to work after only about an hour in the salons. "A French savant," he reported, "even if he made it back to his cabinet, would be tormented by the thought that that evening...he had been gauche in a conversation he had dared to raise against a genuine savant, such as M. Arago."[38]

Arago brought his polarimeter with him into the salons of Paris. He instructed his instrument maker, Soleil, to replace the thin sheets of mica with pieces of gypsum engraved with decorative images. The carving was done such that under ordinary light, nothing could be seen. But under polarized light, brightly colored images emerged showing flowers, butterflies, and even Arago's name surrounded by laurel wreathes. The effect was sensationally popular among the audiences who would sit together and witness the images appear.[39] The complementary colors of the polariscope had themselves become a kind of cultural product circulating within the particular economy of the salon.

The institution of the salon underwent a renaissance in Restoration France.[40] What I want to focus on here are the forms of communication that took place within these social circles. Historians have pointed to the role of the salon as a site of communicative exchange outside of state control, a public space where people could come together in their capacity as private citizens to debate the affairs of the day.[41] Although politics may have dominated the conversations, the claim here is not that the salons were on the first

[36] "Arago to Mathieu, 4 November 1810," Papiers de la famille Arago, Archives Nationales, MI 372.

[37] Arago's biography appeared as part of a series in Paris papers under the pseudonym, "Un homme de rien." Afterwards published in *Sketches of Conspicuous Living Characters of France*, R.M. Walsh, trans. (Philadelphia: Lea & Blanchard, 1841), 281.

[38] Stendhal, "Voyage en France," in *Voyages en France*, V. Del Litto, ed. (Paris: Gallimard, 1992), 462.

[39] Maurice Daumas, *Arago, 1786–1853: la jeunesse de la science* (Paris: Belin, 1987), 84.

[40] Adeline Daumard, "La vie de salon dans la première moitié du XIXe siècle," *Sociabilité et société bourgeoise en France, en Allemagne, et en Suisse, 1750–1850*, Etienne François, ed. (Paris: Recherche sur les civilisations, 1986); Steven Kale, "Women, the Public Sphere, and the Persistence of Salons," *French Historical Studies*, 25.1 (2002): 115–148.

[41] Jurgen Habermas, *The Structural Transformation of the Public Sphere: An Inquiry into a Category of Bourgeois Society*, Thomas Burger, trans. (Cambridge, MA: MIT Press, 1998), 33; Dena Goodman, *Republic of Letters: A Cultural History of the French Enlightenment* (Ithaca, NY: Cornell University Press, 1994); Etienne François and Rolf Reichardt, "Les Formes de sociabilité en France du milieu du XVIIe au milieu du XIXe siècle," *Revue d'histoire moderne et*

order political. Rather, their interest lies in their particular form of sociability. Partisanship was put aside for rational discourse. Salon goers, in theory at least, offered their opinions to one another as equals, and the only reigning authority was critical reason.

One of the people Arago associated with in the drawing rooms of Paris was the painter Léonor Mérimée. The two men knew each other from the École Polytechnique. They taught together there until the Restoration, when their well-known radical sympathies put them on the wrong end of the change in regime. Mérimée's position was suppressed, and he never returned to the school. Arago left his position voluntarily, and was able to come back after a brief absence. They continued to see each other, however, as they traveled in the same social circles and frequented the same salons. It was while at dinner together at the house of their mutual friend, André-Marie Ampère, that Mérimée first mentioned to Arago the ideas of his nephew, Augustin Fresnel, on the subject of light.[42]

Working independently while serving out his engineering position in the provinces, Fresnel had developed an account of diffraction based on the principle of interference between waves of light. Arago responded favorably, and over the next two years Mérimée served as the conduit as Fresnel sent in his scientific works and Arago presented them to the Academy. Arago himself also wrote suggesting works on the subject of light, and in the following months a friendly correspondence arose.[43] Arago eventually arranged for Fresnel to come to Paris. He encouraged the young man to turn to the subject of chromatic polarization, and collaborated with him in polarization experiments.

Fresnel wrote his first account of chromatic polarization in 1816, in which he explained colors as the product of the interference between the two beams of light that become separated in the doubly refracting crystal.[44] He arrived at his own formulae for the intensity of the two emerging beams:

$$\text{Ordinary intensity} = 1/2 + 1/2 \cos 2s \cos 2\pi (e - o)$$
$$\text{Extraordinary intensity} = 1/2 + 1/2 \cos 2s \cos 2\pi (e - o)$$

where e was the number of "extraordinary oscillations," o the number of "ordinary oscillations," and s the principal axis of the analyzing crystal used to observe the images.[45] The expression $(e - o)$ therefore represented the number of oscillations that took place within a distance equal to the difference between the path lengths traversed by the ordinary and extraordinary rays. Fresnel's

contemporaine, 34 (1987): 453–472; Mary Terrall, "Salon, Academy, and Boudoir: Generation and Desire in Maupertuis's Science of Life," *Isis*, 87 (1996): 217–229.

[42] L. Fresnel, editor's note, in Fresnel, *Œuvres*, 1: 6.

[43] Arago, "François Arago à Augustin Fresnel, 12 July 1815," in Fresnel, *Œuvres* 1: 6.

[44] Fresnel, "Mémoire sur l'influence de la polarisation dans l'action que les rayons lumineux exercent les uns sur les autres," in Fresnel, *Œuvres*, 1: 385.

[45] Fresnel provided a more general set of equations that took into account the angle of incidence of the light upon the crystal. The equations given here are for the simplified case of the angle $i = 45°$. They are taken from Fresnel, "Note, extrait du mémoire Sur les couleurs que la polarisation développe dans les lames cristallisées parallèles a l'axe," in Fresnel, *Œuvres*, 1: 537.

equations did not, as Biot's had, specify the tint of the color involved. They did, however, explain the similar patterns between the colors of polarization and Newton's rings. According to Fresnel, the color of the images depended on the particular wavelengths that fit integrally into the distance determined by $(e-o)$ and thus interfered constructively. Colored rings, in a similar manner, were produced by the wavelengths fitting integrally into the thin laminae. To support his claim, Fresnel pointed to how well his equations fit with Biot's data. He insisted, however, that "this was not the case of a simple analogy between the two phenomena."[46] Rather, the same colors were being produced because light was experiencing the same difference in path length in both cases. Biot, on the other hand, "following the system of Newton" had simply assumed the validity of the analogy and rested his entire theory upon this assumption.[47] This, said Fresnel, led to "inconveniences" that his own theory aimed to avoid.[48]

Fresnel had thus succeeded in giving an explanation of the phenomenon of chromatic polarization without Biot's necessary detour through Newton's algorithms for colored rings. His explanation did not, as Biot's had, allow one to come up with a number that could be translated into a particular color. But in this sense it was like his uncle Mérimée's system of color scales: the "color" of the image was what the observer chose to call it, and not something determined by an absolute scale. Fresnel submitted this work to the Academy of Science in 1816. Since he was not a member, the memoir was not presented to the Academy as a whole, but given to a small committee headed by Arago. Their job was to summarize and evaluate the work for the Academy at large, a task which usually took a few weeks. Arago, however, took nearly five years. He was crafting Fresnel's work into an argument for his own position in the debate about polariscopic color. And his principal interlocutor in this debate had meanwhile quit the scene.

Biot's search for order

Unlike Arago, Biot saw in Paris society a swirling mess of danger and instability. As Arago embraced the forms of civic sociability available in France, Biot wrote to an English colleague condemning the "political Experiments" of the past twenty-five years and expressing his desire to "return to those principles consecrated by the history of all time."[49] By 1814 he had tried "a thousand times" to find refuge for himself and his family in England, but had run up against interrupted lines of communication. "I would be far from forming a similar wish if the Bourbons finally come into peaceable possession of the throne," he wrote, but was afraid that the state of "public sentiment" made the

[46] Ibid., 409.
[47] Ibid., 407.
[48] Ibid., 408.
[49] "Biot to Blagden, 1814," Bla.b.155, Archives of the Royal Society of London.

prospect of peace unlikely.[50] If by some ill fate France once again fell prey to *"des orages politiques,"* he inquired, would it be possible for him to find a position with which he could support his family in England? He pointed out that he could teach physics and applied mathematics, and had just completed his manuscript for a textbook on experimental mathematical physics that he would be happy to publish in England. Biot also turned to David Brewster, Britain's leading authority on optics. As they discussed polarization, he also sold Brewster an article on magnetism for Brewster's encyclopedia and asked him to find a publisher for the *Traité.*[51]

Biot did not make it to England in 1814, and he wound up publishing his textbook in France. He kept trying, however, to find a way across the Channel. In 1817, he proposed a new meridian measuring expedition in the British Isles, jointly undertaken with British military engineers. Biot had gone as far north as Dunkirk in his previous measuring efforts. The British had measured from the southernmost point in England to the northernmost point in Scotland. Together, Biot and the English would connect France to England, and extend the meridian line north through the Shetland Islands. The measurement of this arc would give "the most complete, and one may say, the most European, determination of the meter that one could ever hope for."[52] In addition, Biot proposed to perform measurements on the shape of the earth so that the meridian could be corrected for the earth's flattening. He carried with him wherever he went a "seconds pendulum" designed to beat a period of one second. After measuring the length at various locations, he concluded that the weight of a body was not the same at all points on the same latitude. These differences, which he attributed to differences in the thickness of the earth, could be used to correct the geodesic measurements.[53]

Biot arrived in England in the spring of 1817.[54] His description of London was modestly positive: "No more disease than there is in general in any big city; and what little exists is in the inferior classes."[55] Most important, he assured his wife, was the desire for peace that he found everywhere evident. The few worries he had before he left were put to rest when he saw that no one in England had any desire to disrupt the fragile calm. "Yes, yes, yes," he joked in English, "we shall be very well here."[56] Which seems to have been exactly what the Bureau of Longitudes was worried about. Word got back to Biot that the Bureau, after waiting to hear from him for several weeks, had come to suspect that he had taken the chance to emigrate. Biot wrote to Arago to assure him of his intention to complete the mission.[57]

[50] Ibid.

[51] "Biot to Brewster," BI MS489, 61 and 63.

[52] "Biot au Bureau des Longitudes, 4 July 1817," BI MS4896, 111, p. 5.

[53] Biot, "On the Length of the Seconds Pendulum, Observed at Unst, the Most Northern of the Shetland Isles," *Edinburgh Philosophical Journal*, 1 (1819): 77–79.

[54] "Biot to Madame Biot, 11 May 1817," BI MS4896.

[55] "Biot to Madame Biot, 6 December 1817," BI MS4896.

[56] "Biot to Madame Biot, 13 June 1817," BI MS4896.

[57] "Biot to Arago," BI MS4895, 24.

The measurements went slower than expected. In July, Biot wrote to Arago asking him to "come spend two or three months with me with a big repeating circle, bigger than the one I have today."[58] He arranged for Arago to meet him in London, where he could also pay a visit to the Greenwich Observatory. Arago did show up with the instruments. He did not, however, quite last the two or three months. Apparently less sensitive than Biot to the proper social boundaries of disease in Great Britain, he took ill rather quickly and returned home as fast as he could.[59]

As Arago returned to Paris, Biot continued north to the Shetland Islands. He reported his admiration of the austere morality and character of these rural people, living in a profound peace removed from the troubles of the rest of the world.[60] He had plenty of time to admire them, as the English contingent of the expedition accidentally left him behind when the meridian measurements were finished. The families of the island took him in to their homes. After completing his pendulum measurements on his own, he managed to hitch a ride back to Scotland on a merchant vessel.

Biot spent the next several months living among the most distinguished families of Scotland. He maintained a tender correspondence with his wife. He responded with sympathy to her reports over a distressing encounter with the city's underclass. She had been moved by their plight, but, he assured her, she need not feel bad. "Imagine that in this unequal but necessary Distribution of riches, there will always be a poor class."[61] Imagine further, he continued, that these poor had never known anything else and had no sense that their lives could be otherwise. They were far from experiencing the unhappiness she would feel in their place. The "physical jouissance" of her privileged life made her, in effect, a different person from those below her.

The problem of the inferior classes remained on the forefront of Biot's mind during his stay in Scotland. With his measuring duties over, he began to apply himself to a new research project: documenting "the moral state of this country, and the general principle that produces this state."[62] Biot was most struck by the parochial schools of Scotland. He visited them and discussed them with the local gentry. He was very impressed by a system in which "the higher teachers," the professors of divinity, for example, could create a system of producing good schoolmasters.[63] The hierarchical, top-down structure allowed those with particular religious authority to put their own moral standards into place. Biot approved of how well the parochial schools managed class differences. "I would very much enjoy showing my country how the institutions of the inferior classes can not only lead to morality, prudent liberty,

---

[58] "Biot to Arago, 'Lettre confidentielle' 28 july 1817," BI MS4895, 18, p. 2.

[59] "Arago to Brewster," University of Edinburgh archives.

[60] Émile Picard, "La vie et l'oeuvre de Jean-Baptiste Biot," *Eloges et discours académiques* (Paris: Gauthier-Villars, 1931), 238.

[61] "Biot to Madame Biot, 13 June 1817," BI MS4896, 155.

[62] "Biot to Madame Biot, 21 October 1817," BI MS4896, 162.

[63] From a conversation recalled by Thomas Chambers, "Thomas Chambers to Biot," BI MS4895, 101.

and every social virtue, but can also be the result and support of them."[64] It was not particularly French of him to think thus, Biot admitted. "But frankly," he said, he was leaning more and more toward the Scottish life.[65]

Biot's interest in Scottish life was more than theoretical. He began making discreet inquires into the possibility of making his living there. While in Scotland, he found time to "chat about polarization" with David Brewster.[66] Brewster was one of the leading British figures in optics, and was delighted to hear the news of recent French developments. Brewster was also one of the most unconventional practitioners of science in Britain. He worked largely outside the institutional structure centered in Cambridge and the Royal Society, and made most of his money by writing articles on science. Biot saw this career path as a possibility to earn his own keep in the country. Accounts of polarization were in high demand, and few people were better prepared to produce articles on it than he did. He happily reported to his wife that he had already been promised enough work of this sort to allow them to "live very honorably" in Britain.[67]

He was, he explained to his wife, torn about whether to go live in England or not. His only desire was for stability. He would return to France in the case that "everyone desired peace, the reclamations of the press rendered it odious and contemptible, and the demonstrations of resistance made by the king with moderation, but with firmness, render our nation more respectable."[68] Biot's willingness to live in France thus hinged upon the existence of sufficiently compelling forms of central authority. He saw the press, in its capacity of critical opposition, as a force of destabilization. Only by discrediting its arguments through contempt and reestablishing the constancy of the king could France return to the kind of society in which Biot wanted to live.

Biot returned to France in 1818, but resisted settling permanently in Paris. Within a year, he left for Dunkirk on the northern coast to complete the meridian triangulation between France and England with the English team. He showed little desire to hurry home from his location. He spent several months in Dunkirk and integrated himself into the local culture. In 1820, the Société d'Agriculture de l'Arrondissement de Dunkerque named him a honorary member.[69] Once the meridian triangulations were completed, Biot traveled to Spain and Italy to continue collecting pendulum measurements. He stayed away until 1821. Charles X had recently come to the throne in France, promising a renewed commitment to the principles of absolutism and showing no little contempt for the reclamations of the press. Perhaps convinced that this regime would

---

[64] "Biot to Madame Biot, 21 October 1817," BI MS4896, 162.

[65] It is unclear whether Biot intended the pun. "Biot to Madame Biot, 21 October 1817," BI MS4896, 162.

[66] "Biot to Madame Biot, 7 October 1817," BI MS4896, 161.

[67] "Biot to Madame Biot, 6 December, 1817," BI MS4896.

[68] Ibid.

[69] "Société d'Agriculture de l'Arrondissement de Dunkerque to Biot, 20 Jan 1820," BI MS4895, 104.

provide the "firmness" necessary to maintain peace, Biot finally returned to Paris with the intention to stay.

The Battle of 1821

By 1821, Arago had been sitting on Fresnel's memoir for five years. These years had passed in relative peace between Arago and Biot. They had exchanged virtually no words with one another over the subject of chromatic polarization. Yet they both continued to discuss the subject in the diverse locations that they frequented. Arago impressed the audiences of Paris salons with his colorful displays, and Biot explored the possibility of using articles and textbooks to buy his way into a calmer life on the British Isles. These two locations offered strikingly different models of social interaction. The egalitarian critical debate of the salon was precisely the sort of "reclamation" that Biot saw as a threat to the security of France, and wished that a sufficiently firm monarch would keep in check. These communities had different rules about how communication would be regulated, and how the representations that formed the units of communication would circulate. These rules applied to the images of the polarimeter as much as to anything else. When the two men did began to discuss these images again with one another, it was hardly surprising that their disagreement over how people came to agreement over their color remained a live issue.

On June 4, 1821, Arago read his referee report aloud to the Academy of Sciences.[70] His first sentence claimed to show that Biot's explanation of the bright colors of crystal laminae was "insufficient or inexact," precisely the phrase used to launch Arago's initial attack on the theory of fits in 1811. The report only grew more polemical. Arago reported on observations not included in Fresnel's memoir that raised problems for Biot's theory, and generally showed himself to be, as Fresnel later wrote to his brother, "determined to tell the Academy that [my experiments] completely overturned Biot's theory."[71] Where Fresnel had been interested in exploring the power of the principle of interference, Arago had inserted his work into the ongoing debate over how to talk about polarized colors. Arago even changed the title of the paper in his report, from "Memoir on the influence of polarization in the action which luminous rays exercise on one another" to "Memoir relative to the colors produced by doubly refractive crystals."

Biot began protesting as soon as Arago finished speaking. He claimed that the report had been handled inappropriately, and demanded that the Academy reject its conclusions. Laplace postponed the Academy's decision on the report

---

[70] Ampère and Arago, "Rapport fait à l'Académie des Sciences, le lundi 4 Juin 1821 sur un mémoire de M. Fresnel, relatif aux couleurs des lames cristallisées douées de la double réfraction," in Fresnel, *Œuvres*, I: 553–568. Note that the title Arago gave in his report was different from that of the original memoir.

[71] Fresnel, "Lettre d'Augustin Fresnel à son frère Léonor du 13 juin 1821," in Fresnel, *Œuvres*, II: 853; See also Buchwald, *The Rise of the Wave Theory*, 237.

to the next week, and allowed Biot to present a rebuttal. Biot's defense of his work centered around the claim that both systems explained the phenomena equally well, and that his was methodologically more sound.[72] His own formulae, he claimed, were in fact equivalent to those of Fresnel. If one represented the color coefficients as

$$O = \cos^2 \pi/\lambda \, (e - o)$$
$$E = \sin^2 \pi/\lambda \, (e - o)$$

then it was just a question of algebra to move from one to the other. Furthermore, he claimed, his own equations rested upon a firmer foundation. Fresnel, he argued, had relied on crucial assumptions about the nature of light in constructing his equations. He himself had relied on nothing more than the claim that the colors followed the same set of relations Newton described for thin films in the *Opticks*, and this claim was not a hypothesis but a fact meticulously proved by his volumes of experimental data. "I believed, I still believe," he wrote about the relations between the colors of polarized light and those of Newton's rings, "that these are simple physical laws independent of all hypotheses, and to which one cannot with foundation raise objections."[73]

Fresnel's theoretical claims, he pointed out, were not so firmly grounded in years of experimental effort, and needed to meet a more stringent set of requirements. If any of the logical consequences deduced from them could be shown to be wrong, the hypothesis must be rejected. Biot then presented his best case that Fresnel had run into precisely this situation. Fresnel's intensity laws could only actually predict what color would be seen if used in conjunction with something like the Newtonian color circle. That is, Fresnel's laws provided a list of all the various wavelengths involved. But to know what color would result from their combination required a computational method for moving between complex and simple colors. Biot then argued that Fresnel's equations, when taken together with Newton's color circle, gave results that contradicted those of Newton's table of tints.[74] Biot worked through an example. For a thickness of nine thousands of an inch, Fresnel's formulae gave a tint composed of 7/13 red, 2/5 orange, 7/25 yellow, 1/10 green, and 1/16 violet. Using Newton's color circle, one could assign to this an equivalent consisting of a red mixture composed of nine parts white and sixteen parts spectral violet (i.e., the two colors would be indistinguishable to the eye). But, Biot pointed out, this result was in contradiction with the results of Newton's table of tints, which gave the correct color as equivalent to a pure red on the far end of the spectrum.

Both Arago and Fresnel came back with retorts. Arago not only concentrated mostly on defending the propriety of his actions, but also attacked

---

[72] Biot, "Remarques de M. Biot, sur un rapport lu, le 4 Juin 1821, à l'Académie des Sciences, par MM. Arago et Ampère," in Fresnel, *Œuvres*, 1: 569.

[73] Ibid., 1: 574.

[74] Ibid., 1: 578.

the content of Biot's comments.[75] It was a worthless claim, Arago responded, to say that the formulae were equivalent. That implied that one could move from one to the other through purely mathematical transformations.[76] Yet Biot, Arago pointed out, had no way of deducing the numerical values for the color coefficients. Unlike Fresnel, Biot could only obtain these values by reading them off of Newton's table. And that step is what Arago had objected to all along.

Fresnel's remarks went even further.[77] He pointed out that Biot's criticism rested not only on Newton's table of tints, but also on the use of the color circle. Yet this color circle was of decidedly questionable value. Although Biot claimed it reduced a complex mixture of colors to an equivalent spectral color, there were many cases in which it did not work. Fresnel mentioned certain colors, such as those of some flowers, that, even though they were composed of heterogeneous rays and should have lots of white, were still as bright as any spectral color. Some compound colors, such as pink and purple, produced sensations on the eye that had no equivalents in the spectrum. "One thus should regard [Newton's construction] only as a rather rough representation of the quite varied sensations that diverse combinations of rays have upon us."[78] Indeed, it would be hard to assign any exactness to a construction that, "can be so varyingly interpreted by different observers, according to their manner of sensing and naming colors."[79] Fresnel then reported that his uncle, the painter and color theorist Léonor Mérimée, often gave names to colors which were substantially different from those used by Biot. The point was not that Biot was wrong, but that the project of an authoritative color table was flawed. One could only judge color by directly comparing two tints that were side by side, and even that established an identity of sensation and not of composition.

"There has been a great battle," wrote Fresnel in 1821 after the volley of exchanges between Arago and Biot. And historians have tended to agree. Arago's triumph (for indeed, the Academy accepted the conclusions of his report and Biot dropped the subject soon after) has been linked to no less than the abandonment of eighteenth-century Newtonian orthodoxy and the inauguration of a new discipline of physics. Yet this "optical revolution" is nearly always described as a battle between the particle and wave theories of light, which were certainly only peripheral concerns to the central combatants. Arago scarcely mentioned the issue once in the years 1811–1821. Even when delivering favorable reports on Fresnel's memoirs to the Academy of Sciences, he made it clear that he reserved judgment on Fresnel's claims about the nature

---

[75] Arago, who appears not to have fully understood the implications of Biot's comments, also attacks Biot's claims about the azimuth of polarization for thick and thin crystals. Buchwald, *The Rise of the Wave Theory*, 249.

[76] Arago, "Examen des remarques de M. Biot," in Fresnel, *Œuvres*, 1: 591.

[77] Fresnel, "Note sur les remarques de M. Biot," in Fresnel, *Œuvres*, 1: 601.

[78] Ibid.

[79] Ibid., 602.

of light. Biot was likewise only tentative in his support of light particles. In his earliest publications on chromatic polarization, he emphasized that his procedure for determining the color of rays did not depend on his assumptions about the nature of light. By 1821, he was stating very vigorously that he did not care whether light was a particle or a wave, that it made no difference to the formulae he had established, and that any statement on the subject would be premature at that point.

It is telling that the final word on the debate was a discussion of the words Léonor Mérimée used when describing color. The argument had always been at some level on how to talk about color. The technical problem of deriving the expression of the color coefficient was also the practical problem of what to call the colors emerging from the polarimeters that Arago and Biot presented as their cultural products. It was not, ultimately, a question decided only by physicists, as Mérimée's recurring presence attests. Nor was it a question that only interested physicists. The polarimeter, we shall, quickly began infiltrating a wide range of cultural spaces in Restoration Paris. It spanned the world of fine arts and the world of base industry. In each location, the question was posed again and again: how does one talk about the colors that this object produced? The issue of standardization remained a crucial one, and, as before, the issue of whether or not color was immediately perceptible was closely tied to the question of whether communication should occur within or without the bounds of state jurisdiction.

The public use of private color

While Arago and Biot were arguing on the floor of the Academy of Sciences, Mérimée continued to advocate a well-arranged juxtaposition of opposing colors. He provided a section in Gaspard Grégoire's *Théorie des couleurs* that allowed painters to come up with the complement of any given color.[80] The basis of this theory was that all colors could be produced from the three primaries. Merimée and Grégoire produced a table of twenty-four "bright" colors that quantitatively expressed each color as a ratio of yellow, red, and blue. For any given color, it would become a simple matter of arithmetic to find its complement. For example, one could take color #4, marigold, which consisted of 5 yellow, 3 red, and 0 blue. To produce gray, one needed an equal amount of each of the primary colors, and this would be achieved by adding 2 red and 5 blue. This technique gave a precise formula of the mixture needed for the right shade of blue-violet to complement marigold.

The color table could also demonstrate the fundamental nature of the primary colors. If two primary colors were mixed together, they produced a

---

[80] Mérimée, "Théorie de la colorisation appliquée à l'harmonie des couleurs," in Grégoire, *Théorie des couleurs appliquée*, 2nd ed. (Paris: Bachelier, 1939), 11.

secondary. If two secondary colors were mixed together, say violet and green, they produced a color that was 1/4 blue and 3/4 gray.

Violet	0	4	4
Green	4	0	4
	4	4	8

Mérimée's most complete expression of his color theory came in 1830 in *De la peinture á l'huile*.[81] He presented this work as his ultimate statement of the relationship between science and art. Its chief accomplishment was the presentation of thirty years of research on the chemistry of varnishes and pigments. He had undertaken this work while at the École Polytechnique, where he shared laboratory space with many of the chemists involved in the revolt against Newtonian color theory. Mérimée felt, as did many others at the time, that modern technique was in a state of decadence, and contemporary artists were producing shoddy works whose pigments would not retain their color over time.[82] For him, salvation lay in the methods of the early Flemish school. On a trip to Holland in 1789, he had been struck by the excellent state of conservation of the 300-year-old paintings of Van Eyck.[83] He was disappointed to learn that the knowledge of their techniques was lost and, when the resources became available to him in 1795, he set out to try to uncover these "secrets of the old masters." His principal discovery was that they mixed their oil paints with a varnish of dissolved resin.[84] Mérimée revealed this insight, as well as the more recent contributions of research chemists in developing new colors for use by French artists. His own experiments in this regard were significant. Twenty-one pages were devoted to the description of a blood-red madder of his own creation, called *lacque de garance*.[85]

The section that would draw the most renown, however, was separated out at the end of the work. In it, he presented the notions of the color wheel and complementary colors that appeared in his earlier works, then built from new principles a theory of harmony in painting. His stated aim was to bridge physics and art by founding a color theory on scientific principles. He made it clear, however, that by "scientific" he did not mean "Newtonian." Indeed, he ground his project on the anti-Newtonian critique upheld by his fellow professors at the École Polytechnique in the 1790s and continued by Arago. In particular, he rejected the insistence of some physicists that colors must be treated as complex and analyzed into their various components. Mérimée sided with those who treated color as simple: "Regarding them merely as to the sensations which they produce

---

81 Ibid.
82 Gage, *Color and Culture*, 214.
83 Gaston Pinet, *Léonor Mérimée (1757–1836)* (Paris: H. Champion, 1913), 72.
84 Ibid., 79.
85 Léonor Mérimée, *De la peinture à l'huile*, 144–165.

upon the eye, and without reference to their physical properties, I do not find my opinions opposed to those of men of science in this theory of coloring."[86]

The scientific principles Mérimée leaned on were those of chromatic polarization. He had founded his theory upon complementary colors, he said, because nature itself, working through the polarimeter, "points out to us these oppositions."[87] His text contained an extensive description of how a polarimeter produced its colors. He explained the nature of polarized light, how it was produced, and what properties it possessed. He then described how one could make them "display very lively colors and numerous shades of difference, which the natural rays never show."[88] Using mica or certain other thinly cut crystals, one could produce two images, one turning around the other, both perfect complements, as could be seen by the complete absence of color in the areas where they overlapped.

This notion of complementarity, Mérimée claimed, held an important lesson to painting: that mixing diametrically opposed colors produced absolute decoloration. The reason was that opposed colors always represented the union of the three generative colors. If one was simple, the other was a binary composite. They were, as Mérimée said, "always reciprocally complementary."[89] Others before him, he said, had used the term "enemy colors" to designate the set of oppositions in question. It was true, he avowed, that the pairing of these colors formed the strongest opposition that any mixture of the three generative colors could provide. "But these oppositions are only destructive to harmony when they are not prepared; when they are suitably placed, they attract and win over the spectator."[90] Blue and orange in particular had been criticized as discordant, as involving the most intense color and the binary composite of the palest and most brilliant. And authors such as Paul Lommazzo and Lairesse also spoke against using the pairs yellow/violet and red/green. These opinions showed, said Mérimée, that our senses can sometimes be deceived by "vicious habits," and that, when forming principles, one should only accept evidence compatible with physical law and directly induced from experience.[91]

Mérimée's central principle was the following: harmony resulted from a disposition of tones and colors that attracted and fixed the eye by a well-arranged succession of rest and opposition. "Far from destroying harmony," he stated, "the oppositions animate it." A painfully brilliant color became pleasant if placed next to a means of resting the eyes. That was why red, the most brilliant, went well with green, the color that occurred most in nature and which the human eye was most accustomed to seeing. The golden and purple tones of a sunset were attractive to the eye, although more extreme than any of the colors on the artist's palette.

---

[86] Ibid., 244.
[87] Ibid., 248.
[88] Ibid., 249.
[89] Ibid., 275.
[90] Ibid., 288.
[91] Ibid., 289.

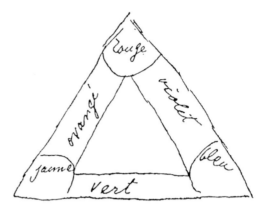

Fig. 2.6 Delacroix's triangle for mixing complementary colors. (From Eugène Delacroix, notebook, Musée Conde, Chantilly, France.)

One artist who took note was Eugène Delacroix. He and Mérimée came into substantial contact during and after the Revolution of 1830, as they served together on a committee to examine the state of the arts.[92] Soon after Delacroix began to experiment with complementary colors.[93] In his notebooks of 1832, he reproduced a version of Mérimée's chromatic circle that emphasized the oppositional arrangement of the primary and secondary colors (see Figure 2.6).

Delacroix also included the following note, "The three secondary colors are formed from the three primary. If you add to a secondary color the primary that is opposite, you will neutralize it; that is to say you will produce its essential half-tone. Thus to add black is not to produce a half-tone; it dirties the color whose true half-tone is to be found in the opposed colors of which we have spoken."[94] His working notebooks, moreover, show frequent reference to the system through such phrases as, "A color for the halftone of gold, nicely suited to accentuating what surrounds it, by opposition."[95]

[92] Delacroix painted Liberty Leading the People, or Les Trois Glorieuses, during the Revolution of 1830. The top-hatted is now generally agreed to be François Arago's brother, Étienne. Lee Johnson, *Historical and Literary Subjects: The Paintings of Eugène Delacroix: A Critical Catalogue, 1816* (Oxford : Clarendon Press, 1981–1989), 143. At least one historian has criticized this presence as overstating the actual involvement of the middle and upper bourgeois in the fighting, David Pinkney, *The French Revolution of 1830* (Princeton: Princeton University Press, 1972), 255.

[93] Gage, *Color and Culture*.

[94] Eugène Delacroix, *Voyage au Maroc, 1832: Lettres, Aquarelles et Dessins* (Paris: Beauxarts, 1930).

[95] Delacroix, *The Journal of Eugène Delacroix*, Walter Pach, trans. (New York: Crown Publishers, 1937), 245.

Delacroix' novel coloring caught the eye of the critics. Prosper Mérimée, son of the artist, wrote a review of his "Cleopatra and the peasant" in the Salon of 1839. The most noticeable feature, he wrote, was the "sudden and hard opposition offered by the orange cloth covering Cleopatra's breast and the blue cloak draped on her knees."[96]

Baudelaire also homed in on the imposing effect of Delacroix's hard oppositions. In the *Salon de 1845*, Baudelaire praised the *Last Words of Marcus Aurelius* as one of the most complete works the art world had to offer. Its color, he claimed, drew its force from "an incomparable science."

> ...the color, far from losing its cruel originality in this new and more complete science, is even bloodier and more terrible.—This heaviness of green and red pleases our spirit.[97]

Baudelaire reminded his readers that coloring was no easy affair. It was hard enough to get the shading right when dealing with a single tone. And when painting in color one had to get the tint right as well, which involved, as science had recently shown, an understanding of the complementary colors at play in all shadows.

> [One must] first find the logic of the shadow and light, and then the correctness and harmony of tone; in other words, if the shadow is green and a light red, one must find on the first try a harmony of green and red, the one dark, the other light, which gives the effect of a monochromatic and *turning* object.[98]

The object in question was in all likelihood a polarimeter. During demonstration events, as the instrument was made to run through its colors by rotating the crystal, the extraordinary image would turn in a circle around the ordinary one. Baudelaire may have seen one for himself in the salons of Paris or read Mérimée's lengthy description, which mentioned its turning nature.[99]

Another art critic, writing a review of the Salon of 1824 under the name M. Van Eube de Molkirk, remarked upon Delacroix' "feeling for color."[100] His review of most of the works of that Salon was a rather scathing pan. He found that most paintings failed to convey what he called "the character of the painter, his manner of feeling the events of his life."[101] They lacked "that individuality which leaves a strong impression."[102] Delacroix stood apart from the rest, in his eyes, because he alone "dared to be himself, at the

---

[96] Prosper Mérimée, *Journal des débats* (1839).

[97] Baudelaire, *Le Salon de 1845*, André Ferran, ed. (Toulouse: Editions de l'Archer, 1933), 146.

[98] Ibid.

[99] Georges Roque, *Art et science de la couleur: Chevreul et les peintres de Delacroix à l'abstraction* (Nimes: J. Chambon, 1997).

[100] Stendhal, "Critique amère du Salon de 1824 par M. Van Eube de Molkirk," in Stendhal, *Œuvres complètes*, 47: 39.

[101] Ibid., 39.

[102] Ibid.

risk of being nothing."[103] And thus the immediacy of color, its transparency as representation, stood as this critic's requirement for participation in the Salon economy.

In 1830, Van Eube de Molkirk would switch to the less ridiculous pen-name Stendhal and leave art criticism for fiction. In 1834, he penned *The Pink and the Green*. This novel, which remained unfinished, set the charming innocence of a young heiress against an appalling backdrop of greed and manipulation. Stendhal reported that he had been inspired by the sight of a young girl sitting across from him on a train. The greenness of her bonnet, he claimed, served to bring out the complementary shade of red in her cheeks.[104] Where Stendhal learned about complementary colors is not clear, and he may or may not have first witnessed the effects in a salon. But his participation in these circles was renowned. He was known as the inseparable companion of Delacroix and Prosper Mérimée at the salons of Paris. He spoke of his encounters there with Arago, the only man of science to receive his consistent praise.[105]

Yet artists were not the only ones producing colors. Outside the drawing rooms of the salons, a growing industrial dye sector was churning out brightly colored fabrics for an expanding market. To this end, I want to investigate the ways in which the growth of industry, or at least state-run, "organized" industry, was itself tied to a process of social differentiation within the communicative sphere. The colonization of industry, particularly the communication industry, by the state is often characterized by a movement from a sphere of private exchange in which as many people expressed ideas as received them, to a divided elite and mass audience, where far more people received ideas than expressed them.[106] The production of this representation is no longer a process guided by individual judgment. It has become part of a commercial enterprise. An intrinsic part of this enterprise became the increasing separation between the production and consumption of representations.

---

[103] Ibid.

[104] Stendhal, "Introduction," *Le rose et le vert: Mina de Vanghel; suivis de Tamira Wanghen et autres fragments inédits*, Jean-Jacques Labia, ed. (Paris: Flammarion, 1998).

[105] Stendhal's library contained several of Arago's works, and in particular copies of the Bureau des Longitudes' *Annuaire* that Arago edited. When Stendhal once wrote a letter to Arago inquiring about the effects of full moon on vegetation, he signed his name, "H. Beyle. Assiduous reader of *l'Annuaire*." Stendhal, *Correspondance*, III: 200–201. The virtually unique admiration of Stendhal for Arago among men of science is detailed in J. Théodoridès, "Stendhal et Arago," *Reflets du Roussilon*, 35 (1985); J. Théodoridès, *Stendhal du Côté de la Science* (Lausanne: Editions du Grand-Chêne, 1972), 266–267. Stendhal also knew of Biot. He seems to have learned much of his physics from Biot's *Traité de Physique*, in which Biot presented his strongest case for the Newtonian color project. He mentions the text several times in his correspondence, Stendhal, *Correspondance*, I: 859; II: 270; I: 1091. Stendhal was not impressed. If Arago was "a genuine savant," Biot was, again in Stendhal's words, "a charlatan workman." Stendhal, *Vie de Henri Brulard* (Paris: H. et E. Champion, 1913), 2: 36.

[106] Jon Cowan, "Habermas and French History: The Public Sphere and the Problem of Political Legitimacy," *French History*, 13 (1999): 134–169.

Dye industry as state enterprise

The project of standardization, the close alliance of state, science, and industry, and the use of complementary colors would all come together in the work of Michel Eugène Chevreul. Chevreul had worked on the coloring principles of dyestuffs since he first came to Paris to seek his fortune as a chemist in 1803. He had started as an unofficial assistant to Nicolas Vauquelin, and accompanied him to the Muséum in 1804.[107] Vauquelin's own research involved using solvents to extract organic materials of medicinal use from plants and animals. He set Chevreul to work extracting materials used in the dye industry. Chevreul's earliest work was on pastels from wood.[108] He threw himself into the study of indigo in 1806, when the Continental Blockade cut off supply of this dye from the West Indies.[109]

In 1824, the state put Chevreul in charge of overseeing the dying of cloth at the national factories of Gobelins (tapestry), Beauvais (upholstery), and La Savonnerie (carpets).[110] "Les manufactures," as they were called, were products of seventeenth-century mercantilist efforts to centralize production under a state monopoly. Their products dominated the market for luxury cloth and set the standard in high-end taste. Napoleon placed the first chemist in charge of the Gobelins dye workshop in 1803, and the entire factory devoted considerable resources to solving the blockade's indigo crisis.[111] Chevreul was, however, by far the best-known chemist to be given the position.[112]

Soon after his arrival at Gobelins, Chevreul began giving a series of courses there dedicated to the application of chemistry to dyeing.[113] The course began with an overview of basic chemistry that drew heavily from Berthollet's

---

[107] It was in this capacity that Chevreul first met Mérimée. Before turning to optics, Fresnel had tried his hand at a new method of producing soda from sea salt. In 1811, he wrote to his uncle, who then asked the eminent chemist Vauquelin to verify the young man's results. Vauquelin, himself very busy, assigned the project to Chevreul, who was then his student. Mérimée wrote assuring his nephew, saying, "I will go see this student and press him to get busy on [the task]." Mérimée, "Lettre de Léonor Mérimée à son neveu Augustin Fresnel," in Fresnel, *Œuvres*, 1: 813.

[108] Chevreul, "Expériences chimiques sur les bois de Brésil et de Campêche," *Annales de chimie*, 66 (1808): 225–265; Chevreul, "Recherches chimiques sur le bois de campêche et sur la nature de son principe colorant," *Annales du muséum d'histoire naturelle*, 17 (1811): 280–309.

[109] Yves Laissus, "Un chimiste hors du commun: M.-E. Chevreul (1786–1889)," *Catalogue de l'exposition sublime indigo* (Marseille: Musées de Marseille, 1987); Jenny Balfou-Paul, *Indigo* (London: British University Press, 1998).

[110] Biot had known the previous director of manufactory at Beauvais, and in fact had married his daughter. H. Havard and M. Vachon, *Les manufactures nationales: Gobelins, Savonnerie, Sèvres et Beauvais* (Paris: G. Decaux, 1889); Jules Guiffrey, *Les manufactures nationales de tapisseries. Les Gobelins et Beauvais* (Paris: H. Laurens, 1907); Chantal Gastinel-Coural, *La manufacture de Beauvais du consulat à la IIe République: Tapisseries, cartons, maquettes* (Paris: Administration générale du mobilier national et des Manufactures des Gobelins, de Beauvais et de la Savonnerie, 1998).

[111] Chantal Gastinal-Coural, "Chevreul à la manufacture des Gobelins," in *Michel-Eugène Chevreul: Un savant, des couleurs!*, Georges Roque, Bernard Bodo, Françoise Viénot, eds. (Paris: Editions du Muséum National d'Histoire Naturelle, 1997), 69; F. Caron, "Science chimique et technique chimique en France au début du XIXe siècle," *L'actualité chimique*, (1987): 31–35.

[112] Gastinal-Coural, "Chevreul à la manufacture des Gobelins," 70.

[113] Chevreul, *Leçons de chimie appliqué à la teinture* (Paris: Pichon et Didier, 1830), 2 vols.

affinity theory. Then Chevreul turned to the industrial applications. For each coloring agent, he ran through a discussion of the various forms it had in nature, and how one prepared it for commercial use. He also touched on the means of evaluating the various dyes produced.

The industrial dyer's primary concern was the "venal value" of a coloring material. Chevreul described various ways of evaluating a color's appearance in his *Leçons*, but seemed dissatisfied with all of them. He was particularly critical of the use of a standard colorimeter. In this technique, one dissolved the samples under examination in a solution of sulfuric acid, and then added water to each one until they all had the same depth of color. The idea was that the more water one had to add, the higher the value of the coloring material. One compared each sample to a control by placing it in a colorimeter: a black box with two holes allowing one to compare the light coming through two test tubes filled with liquid. Yet the fault of this procedure, Chevreul pointed out, was that it measured differences in tone without taking differences of hue into account. Commercial indigo, for example, usually had some amount of yellow material mixed in with the coloring agent of indigotine. It would thus have a greenish tinge that would make it impossible to compare easily with a control of pure indigotine sulfate.[114]

The issue of evaluating color was a crucial one for Chevreul's industrial duties. His principal mission upon entering Gobelins was to address complaints that pieces of cloth were losing their dye after leaving the factory. Although he acknowledged that in some cases the dye may have faded, he insisted that often the cloth had not changed color at all. What had changed were the viewing conditions. A satisfactorily dyed black cloth would begin to look faded if placed next to other dark colors after leaving the factory. Chevreul then realized that his duties at Gobelins were twofold: both producing the brightest, most permanent colors possible, and teaching people how to appreciate them.

Chevreul began his program of visual re-education in 1828, and published his best-known work on the subject in 1839.[115] *On the Law of the Simultaneous Contrast of Colors*, although often celebrated as the introduction of a psychological subjectivity into color viewing, was primarily an effort to achieve universal agreement through rational systematization. The work began with a story about how Gobelins' customers were fooled by their own senses into thinking their cloth had faded, but it ended with the proposal of an improved color circle that would keep them from making that mistake again.

The law of simultaneous contrast rested upon a notion of complementary color. It stated that when two colors were adjacent to one another, each

---

[114] Ibid., lesson 30: 64.
[115] Chevreul, "Author's Preface," *On the law of the simultaneous contrast of colors, and on the set of colored objects considered according to this law in its relations with painting, the tapestries of Gobelins, the furniture upholstery of Beauvais, carpets, mosaic, colored glass, dyestuffs, printing, lighting, building decoration, apparel, and gardening*. Based on the first English edition of 1854 as translated from the first French edition of 1839, *De la loi du contraste simultané des couleurs* (West Chester, PA: Schiffer, 1987).

would appear to be tinged with the complementary color to that of its neighbor. Chevreul acknowledged in the opening lines of *On the Law of Simultaneous Contrast* that complementarity was a fundamentally optical phenomenon, a property better seen in light than pigment. He credited the invention of the polarimeter with spreading the correct idea of complementary colors.[116] The fact that it could so easily demonstrate how red and green light, for example, combined to white was, he said, an "auspicious sanction" to his theory.[117]

Chevreul has often been criticized for not distinguishing between the additive colors of the polariscope and the subtractive colors of his dyestuffs, but he was well aware that colors were complex mixtures. This was precisely the task he had to figure out in coming up with a color nomenclature. His answer was to take the typical color circle and add another dimension to it, making a three-dimensional "color hemisphere." The hemisphere had as its base a color circle, with the various pure tints running around the circumference. The inward-pointing axis was the scale of different tones, describing how much white was mixed in with a color. Chevreul then added another quadrant rising up out of the plane of the circle, to describe how much a given color had been "broken" by the addition of black. This extra quadrant was what allowed him to better account for colors as mixtures, since when complementarily colored dyes mix they produce black.

The colors of the hemisphere would be taken from optical instruments. One must, said Chevreul, start with the invariable colors produced by the solar spectrum, polarized light, or Newton's rings. In all three cases, light ran continuously through the full range of pure tints. For each color produced, Chevreul proposed to imitate it as exactly as possible using colored materials. The use of these constant colors ensured the universality of the hemisphere. But it also involved loaded assumptions about the colors of polarized light. A central point of contention in the debate between Arago and Biot had been whether the colors of polarized light could be translated into pure spectral colors by using Newtonian algorithms. Biot had rested on that principle when he proposed his own system for using the polarimeter to standardize colors. Chevreul followed Biot in the matter, ignoring the objections of Arago, Fresnel, and Mérimée that the reduction to the colors represented by Newton's color circle (and thus by his table of tints) was unwarranted.

Once the hemisphere was made, Chevreul asked, what must happen for it to be useful? "It is to be put in practice everywhere, so as to render the language uniform." The hemisphere would function, he claimed, in the same way as a thermometer. Whenever one wanted to take the "reading" of a color, one would turn to the hemisphere's quantified scale. Just as a national standard of the thermometer existed somewhere in a Paris archive, Chevreul made it clear that a physical model of his color hemisphere deserved a spot at the Bureau of Weights and Measures, as well.

[116] Chevreul, *Complément des études sur la vision des couleurs* (Paris: Firmin-Didot, 1879).
[117] Ibid.

He almost got his wish in 1842. In that year, Chevreul went down to Lyon to give his course on dye chemistry to an audience of textile manufacturers.[118] The local Society of Agriculture, Natural History and the Useful Arts then suggested to Chevreul that they write to the Lyon chamber of commerce and ask the minister of agriculture and commerce to establish a physical standard of the chromatic-hemispheric construction for use by the textile industry. The minister approved a project for the Manufacture de Sèvres to construct a hemisphere out of porcelain. Chevreul began pulling together wool samples to send to Sèvres as the "color types" for the construction. In the several years he took to do this, the project fell through, and no porcelain hemisphere was ever made. For Chevreul, however, the wool samples themselves were a sufficient archive. By 1861, he had a physical sample for a good portion of the 14,440 colors of the hemisphere. He published a description of his "master color thermometer" in a text entitled *Exposé d'un Moyen de définer et de nommer les couleurs d'après une méthode précise et expérimentale*.[119]

As promised, Chevreul ensured the absolute nature of his colors by comparing them to optically generated tints. He started out using a polarimeter. As Biot had pointed out, this instrument was particularly suitable for quantification because a color could be precisely characterized by the angle through which the analyzing crystal was rotated. By the time Chevreul began publishing on this project in 1861, he had switched to using primarily solar spectra. Improved techniques had rendered the black Fraunhofer lines increasingly visible, and these served even more conveniently as a way to refer precisely to particular colors. Chevreul nonetheless provided a description of the polarimeter he had used.[120] He worked with Edmond Becquerel, a former student of Biot's at the Faculty of Sciences, using a device consisting of a doubly refracting crystal, a quartz lamina, and a Nichol prism. A diaphragm allowed only one of the two images to pass through. This image (or alternatively part the solar spectrum) was projected on a screen. Chevreul would hold a piece of wool next to it, and he and Edmond Becquerel would decide whether the two were of the same color.[121]

The second-half of Chevreul's project was to show how the hemisphere could then be used to define and name not just wool swatches, but the color of any object. The classification systems of naturalists such as Mirbel, he pointed out, were in desperate need of his rational system of color naming. Obligingly,

---

[118] Lyon was the center of the silk industry in France. Robert Bezucha, *The Lyon Uprising of 1834: Social and Political Conflict in the Early July Monarchy* (Cambridge, MA: Harvard University Press, 1974). The Chamber of Commerce of Lyon also paid for the printing of some of Chevreul's work, as seen in the subtitle of Chevreul, *Théorie des effets optiques que présentent les étoffes de soie, ouvrage imprimé aux frais de la Chambre de commerce de Lyon* (Paris: Didot frères, 1846).

[119] The full title was much longer. Chevreul, "Exposé d'un moyen de définer et de nommer les couleurs d'après une méthode précise et expérimentale avec l'application de ce moyen à la définition et à la dénomination des couleurs d'un grand nombre de corps naturels et de produits artificiels," *MAS*, 33 (1861): 1–933.

[120] Ibid., 92.

[121] Ibid., 35.

he provided the hemispheric values of hundreds of species of minerals, flowers, leaves, woods, and more. He then moved on to the animal kingdom, and properly named the colors of a host of mammals, birds, on down to the zoophytes.

He included the human species in his color classification scheme. Chevreul examined the colorings of five individuals, three from the Latin family, one from the Arab family, and one sub-Saharan African.[122] For each, he provided a set of eight readings of the hair, eyes (three separate parts), lips, cheek, forehead, and arms. The 8-year-old blonde girl, for example, was found to have an arm color of three violet-red 1.5 tone, while the 44-year-old African man had one of four red-orange 9/10 14 tone.[123] The crucial factor separating them was not so much the tint or their skins (although these were different), as it was the "brokenness" factor. The African's reading of 9/10 meant that the basic tint of his skin had been mixed with nine parts black. The blonde, with no number for brokenness, had no black mixed in with her own color. These examples, although not distinguished by Chevreul from the hundreds of others he provided, recall France's place within a large and expanding empire. The color of one's skin was of undeniable importance at this moment, when the condition of slavery in the colonies was a pressing political point.[124] (Arago and Biot had their own roles to play in the colonial saga, as we shall see in Chapter 6, and the issues of transparency and agreement would color their attitudes toward human freedom).

Chevreul ended *De la loi* with an extended discussion of the implications of the law of simultaneous contrast for the possibility of civil discourse. Both, he claimed, were founded upon an act of judgment. The brain "perceives and judges ideas as it judges colors."[125] That is, by comparing and attaching value. Yet the example of color showed just how difficult this task could be. Imagine five people, he suggested, who all saw the same color of red. Then imagine they each saw it again separately under different conditions. If a discussion arose among them, they would never be able to agree. The one who saw the red against a blue background would claim it took on an orangish hue, whereas the one who saw it against a yellow background would claim it now looked more purple. The observers would all be accurately reporting what they saw. "Each would," Chevreul pointed out, "have cause for maintaining his opinion."[126] Yet clearly no agreement was possible. This should be a lesson, Chevreul warned, to anyone who placed too much faith in the possibility of working things out through reason. Man wanted a complete knowledge of all things. But "the feebleness of his nature" forced him to focus on one thing at a time. Each person would see the world, as it were, against a different background.

---

[122] Ibid., 844.

[123] Ibid.

[124] For the French academic discussion of skin color and ethnicity, see Martin S. Staum, *Labeling People: French Scholars on Society, Race, and Empire, 1815–1848* (Montreal/Kingston/London: McGill-Queen's University Press, 2003).

[125] Chevreul, *On the law.*

[126] Ibid., paragraph 1010.

Conclusion

Mérimée and Chevreul have been recognized as the heralds of "a new paradigm" of complementary colors.[127] Their color theories broke with the past by exploring the harmonies of sharp contrasts. Both cited the images of the polarimeter as evidence that nature itself operated through oppositions. Drawing from the physics of polarized light, each devised a color wheel upon the premise that colors opposite one another should be complements.

But despite the similarities, these two color circles were entirely different objects. Mérimée's circle was a private object independently calibrated by each artist. Chevreul intended his circle as a national standard. Distinguishing them was the use of the optical color to root Chevreul's colors in absolute constants. Chevreul's use of the polarimeter to calibrate his circle took up, in a tangible sense, the debate between Arago and Biot over how to use this instrument. Biot had suggested almost immediately that the polarimeter be used to standardize colors. Arago had balked, rejecting the claim, that the complex mixtures that were the natural colors of the world could be so easily reduced to spectral hues. Such an equivalence implied that a person could not, without recourse to Newtonian analysis, know what it was they were seeing. Biot accepted this implication, but argued that it made the project of standardization that much more important: it allowed people to talk about colors even if they did not know what they were.

Historians of science have focused on standardization as a means by which local knowledge is made global. Work focusing particularly on the latter half of the nineteenth century, moreover, has uncovered its crucial role in the project of state building.[128] It may thus seem counterintuitive that Biot, with his far more restricted communicative community, should be the one pushing hard for standardization. But much of this can be explained by recalling that standardization, which sought to locate the authority of a system of signs within the institutions of the state, would arise most forcibly as a project once the ideal of freely mediated communication between private individuals had been destroyed. The salon members viewing the various colors of Arago's polariscopic flowers had no need to coordinate their experiences by looking up the numerical value of a color in Newton's tables. The greens and reds they saw were, by Arago's account, the real greens and reds there, and all that was needed to talk about them was to exercise their judgment. Total transparency

[127] George Roque, "Les couleurs complémentaires: Un nouveau paradigme," *Revue d'histoire des sciences et de leurs applications*, 47 (1994): 405–433.

[128] Michael Gordin, "Making Newtons: Mendeleev, Metrology, and the Chemical Ether," *Ambix* 45, no. 2 (1998): 96–115; Simon Schaffer, "Late Victorian Metrology and Its Instrumentation: A Manufactory of Ohms," *Invisible Connections: Instruments, Institutions, and Science*, Robert Bud and Susan E. Cozzens, eds. (Bellingham, Washington: SPIE Optical Engineering Press, 1991), 23–56; Theodore Porter, "Objectivity as Standardization: The Rhetoric of Impersonality in Measurement, Statistics, and Cost-Benefit Analysis," *Annals of Scholarship*, 9 (1992): 19–59; M. Norton Wise, "Precision: Agent of Unity and Product of Agreement. Part I—Traveling," *The Values of Precision*, M. Norton Wise, ed. (Princeton: Princeton University Press, 1995).

alleviated the need of coordination. Likewise, it would be those with the least faith in rational communication that argued most forcefully for standardization. Biot and Chevreul insisted that it was because people could not rely on their own opinions that they needed the state to tell them what was what.

Differences in sociability were driving Arago and Biot apart personally, as well. One well-known anecdote highlights their varying sense of how open one should be with one's ideas. As Arago and Biot left the Bureau of Longitudes together one day, the conversation turned to an idea Arago had for using the polarimeter as a photometer.[129] Trying to persuade Biot, Arago stopped before the church Saint-Jacques-du-Haut-Pas and sketched a diagram of the instrument with a key upon one of the columns. With Biot convinced, the two men continued on. The surprise came next Monday, when Biot announced to the Academy of Sciences that he had invented a photometer, and proceeded to show the diagram that Arago had just shown him. Arago, in the audience, interrupted to point out that the idea had been stolen, but Biot ignored him. At a loss, Arago reminded him of the drawing on the church column, and asked two Academy officials to be sent to verify its existence. Biot, posterity recounts, did not wait for the officials to return.

But Biot also had reason to complain. He blamed Arago's excessive volubility for his loss in the Academy of Science's 1822 elections. The death of Jean-Baptiste Joseph Delambre had left the seat of perpetual secretary for the mathematical sciences open. This seat (together with the perpetual secretary for the natural sciences) was the true power of the Academy, as the post of President was largely ceremonial and rotated yearly. Although the nominating committee usually suggested a candidate, in 1822 they refused to do so, and instead sent back the list of the candidates in descending order of age (Fourier, Biot, and Arago).[130] Arago then gave a speech before the Academy withdrawing his name. He already had too many appointments, he claimed, and was too busy to devote sufficient time to the job.[131] Before he sat down, however, he added that Biot shared an identical number of appointments, with the clear implication that Biot should withdraw as well. Biot declined to do so. When he lost the subsequent election to Fourier, he held Arago responsible.

---

[129] Daumas, 149. Daumas admits that the date of the anecdote is unknown, and the Procès-verbaux contain no record of it. But his comment that, following the episode, Biot stayed away from the Academy for two years implies that it took place in the 1820s.

[130] *Procès-Verbaux de l'Académie des sciences*, vol. 8 (1820–1822), 386.

[131] Maurice Crosland, *Science Under Control: The French Academy of Sciences 1795–1914* (Cambridge: Cambridge University Press, 1992), 99.

3

Astronomy: The Light of the Heavens

＝＞◦◦◦＜＝

Biot and Arago had first met at the Paris Observatory while making astronomical observations in 1806. Fifteen years later, they were barely speaking to one another and an unlikely pair for collaboration. But they both retained an interest in the heavens. As their public brawl over chromatic polarization drew to a close, each one spent more and more time on astronomical matters. Their paths crossed little, and their work had almost no overlap. But they nonetheless managed to disagree profoundly.

Arago remained at the Observatory, and during the 1820s was placed in charge of observations. He ran a tight ship, increasing the number of observers and purchasing several impressive new instruments. But his greatest renown came from his efforts at public instruction. Hailed as the public face of astronomy in France, Arago devoted enormous effort to debunking common superstitions about the influence of the skies. His campaign had one dominant message: the heavens exerted no mysterious influence over Earth. He calmed the public's fears over the potential effects of passing comets, and dismissed the practice of astrology as an "absurd" relic of the past.

Biot found himself less welcome at the Observatory in the 1820s. But his burgeoning interest in the heavens had little need for telescopes. Fascinated by an ancient zodiac unearthed in Egypt, Biot threw himself into the study of the astronomical systems of pre-Christian civilizations. Virtually founding the field of archeoastronomy, he developed a range of theories about Egyptian, Chinese, and Indian astronomy. Both astronomers and classicists found his theories strange and idiosyncratic. More than just a hobby, Biot considered the study of ancient astronomy the key to unlocking profound secrets of the earliest moments of human civilization. He worked to establish a chronology, reaching back to these first peoples. Their astronomy, he maintained, constituted an advanced form of knowledge where astronomical and religious truth were bound tightly together.

If Arago urged the French public to set aside astrology as superstition, Biot worked to reinstate it to its rightful place in ancient systems. He insisted that in the earliest civilizations, the science of astronomy and the astrological precepts governing behavior were inseparable. Separating them after the fact obscured the knowledge they contained. The zodiac represented a form of pure original knowledge where the laws of the heavens and the laws of man coincided. His interests soon expanded beyond Egypt to include Chinese and Indian zodiacs as well. These ancient zodiacs were a way of getting back to pure original knowledge, before the astrological was corrupted with idolatry.

There is also another sense in which their relationship with the heavens diverged. This period marked Biot's return to the Catholic faith, and his increasing association with Jesuit circles of the extreme right. Arago, meanwhile, remained a famous skeptic. In the words of Victor Hugo, "Arago was a great astronomer. Remarkable thing: he spent all his time looking at the heavens and did not believe in God."[1] God was, of course, little invoked by either camp in their treatment of celestial objects. And yet a fundamentally theological issue, the mystery of the world versus its legibility, provided the framework for their claims about the stars.

Biot's move to the country

In 1822, Biot purchased the estate of Nointel in the region of Beauvais. After his humiliating defeat, so the story goes, he left Paris for the countryside, too ashamed to show his face at the Academy. But Biot was not the only one returning to the countryside in the 1820s. These years saw a widespread rush of the French nobility back to the country property they had abandoned during the Revolution, with shame and defeat the furthest emotions from their mind. The restitution of émigré property was one of the highest priorities of the ultra-royalist government that came to power on December 15, 1821. As part of the effort to restore the traditional order in France, the Minister of Finances, Jean-Baptiste de Villèle, allotted over a billion francs for the purpose of allowing émigrés to buy back the estates confiscated during the revolution.[2]

Biot's own move was hardly a restitution of family privilege. He was born and raised in Paris, the son of a mid-level functionary. His first contact with Beauvais came in 1795, when he left Paris to teach physics at the newly instituted Ecole Centrale de l'Oise. It was there that he met and fell in love with Gabrielle Brisson, whom he married in 1797. The family was a prominent one. Her father, Antoine François Brisson, directed the national upholstery factory that gave Beauvais its renown. Biot had thus married into the local notability, and set up house in a neighboring village. His flight from the city can be read as part of a process so old and widespread as to be inseparable from the makeup of the countryside itself. A fair portion of the *noblesse campagnarde* of the region had started out as urban bourgeoisie. In a process both economic and cultural, indebted manors fell into the hands of rich merchants, who used their estate to found a new life of aristocratic leisure.[3]

[1] Victor Hugo, *Choses Vues 1830–1871* (Paris: Le cercle du livre de France, 1951), 155 (1847).
[2] André Jardin and André-Jean Tudesq, *Restoration and Reaction, 1815–1848*, trans. Elborg Forster (Cambridge: Cambridge University Press, 1988), 60.
[3] Pierre Goubert, *Beauvais et le Beauvaisis de 1600 à 1730* (Paris: École Pratique des Hautes Études, 1960).

Biot spent the next seventeen years largely at his country estate in Nointel. Soon after moving, he took to signing his name "Biot, *propriétaire*" identifying himself as a large landowner. That same year Biot was elected mayor of the town of Nointel. His duties were not vast. They included settling local disputes and offering advice. In general, the position of mayor played a key role in the nobility's efforts to reassert local control over the provincial hamlets of France.[4] For Biot, it marked his participation in the period's return to conservative politics.

Biot's life in the country paralleled another feature of an increasingly conservative Restoration France. The power of the Church grew substantially in these years, particularly after the death of Louis XVIII and the assumption of power of Charles X in 1824. And Biot's religious faith grew as well. As a student of Laplace, he had done little to disturb the equanimity of the secularly inclined group surrounding him. After 1822, however, his move to the country coincided with a return to the fold. Although he made few public statements of faith, he gained a reputation among his colleagues as being profoundly religious.[5] He also began to move in Jesuit circles. In 1825, he traveled to the Vatican to visit the Pope but was almost denied admittance for his suspiciously frequent visits to the Jesuit College in Rome.[6]

The Jesuits were a mysterious and divisive force throughout the Restoration. The Society, shut down by papal condemnation since 1774, was reestablished with the Bourbon dynasty in 1814. It achieved its greatest prominence of the early century in the 1820s, although it was an oddly shrouded form of prominence that involved little public presence. The official number of Jesuits remained under 500 during the Restoration, with virtually all members either describing themselves euphemistically as "Pères de la Foi" or trying to pass as ordinary clergy.[7] The eight Jesuit schools in operation disguised themselves as "diocesan petites séminaires."[8]

But if the Jesuits kept their profile low, they came nonetheless to serve as a lightening-rod of anticlerical criticism. Indeed, the cloak of secrecy only fueled rumors that the Society was masterminding some sinister right-wing conspiracy.[9] The very name Jesuit came to stand for reactionary fanaticism.

---

[4] François Furet, *La Révolution: De Turgot à Jules Ferry: 1770–1880* (Paris: Hachette, 1988).

[5] Emile Picard, "La vie et l'œuvre de Jean-Baptiste Biot," *Éloges et discours académiques* (Paris: Gauthiers-Villars, 1931).

[6] Biot, "Une conversation au Vatican," *Mélanges scientifiques et littéraires,* vol. 2 (Paris: Michel Lévy Frères, 1858), 451–459.

[7] Geoffrey Cubitt, *The Jesuit Myth: Conspiracy Theory and Politics in Nineteenth-Century France* (Oxford: Clarendon Press, 1993). For more on the Jesuits in the Nineteenth Century, see Joseph Burnichon, *La Compagnie de Jésus en France: Histoire d'un siècle, 1814–1914,* 4 vols. (Paris, 1914–22); Auguste Carayon, *Bibliographie historique de la Compagnie de Jésus, ou catalogue des ouvrages relatifs à l'histoire des Jésuites depuis leur origine jusqu'à nos jours* (Paris, 1864); Paul Dudon, "La Résurrection de la Compagnie de Jésus (1773–1814)," *Revue des questions historiques,* 133 (1939): 21–59; Ralph Gibson, *A Social History of French Catholicism, 1789–1914* (New York: Routledge, 1989); Antonin Lirac, *Les jésuites et la liberté religieuse sous la Restauration* (Paris: Palmé, 1879); David Mitchell, *The Jesuits: A History* (London: Macdonald, 1980); John Padberg, *Colleges in Controversy: The Jesuit Schools in France from Revival to Suppression, 1815–1880* (Cambridge, MA: Harvard University Press, 1969).

[8] Cubitt, *The Jesuit Myth,* 20.

[9] Ibid.

As Michelet pointed out in his work *Des jésuites*, every man on the street equated the group with the Counter-Revolution.[10]

The Jesuit menace seemed particularly frightening under the regime of Charles X. The regime's renewed alliance between Throne and Altar was visible from its inauguration. When Charles X assumed the crown in 1824, he did so with the full ritual of the sacré. Intended to cement the theological foundations of the legitimate monarchy, the sacrament/coronation offered the sight, disturbing to many French, of the king fully prostrate on the ground at the feet of Catholic priests. The flurry of conservative legislation that followed the regime's inauguration (and not least the indemnity of émigrés) provoked equal consternation and suspicion of Jesuit involvement.

In Beauvais, Biot became a member of the inner circle of the controversial Jesuit extremist, Gustave Xavier Lacroix de Ravignan.[11] The figure of Père Ravignan was a focal point of Jesuit suspicion.[12] He was one of the only Jesuits to admit he was a member of the Society of Jesus or speak openly of its existence. Roughly the same age as Biot, he had also left a promising career in Paris in 1822. After training as a Jesuit, he moved to the countryside near Beauvais to proselytize. He spoke fondly of the "religious sentiment" of the countryside.[13] Yet like Biot, he found himself the object of rural animosity in 1830, when the peasants both ransacked his home and threw rocks at his person.[14]

Ravignan stressed the fundamental mystery at the center of human knowledge. He was not opposed to the practice of science. "Advance with courage, o you zealous propagators of science," he encouraged in one of his celebrated addresses at the Notre Dame de Paris.[15] The noble faculties of reason allowed man to begin with his observations and proceed to deduce new truths by establishing relations between already secure knowledge. Reason thus advanced into the vast fields of the unknown, Ravignan recounted, until "all at once, its sight grows dark."[16]

> Messieurs, far from the intelligent eye of the soul, outside the limit of its ideas and natural experiences, there still lie vast regions of truth. For beyond that is the invisible, the incomprehensible, we cannot doubt it. God with his intimate essence lives in the inaccessible light.[17]

[10] Jules Michelet, "Des jésuites," in *Des jésuites*, J. Michelet and E. Quinet, eds., 7th edn. (Paris: Hachette, 1845), 22.

[11] A. de Lapparent, "Biot," *Biographie centennaire*, Archives of the Ecole Polytechnique, 261; F. Lefort, "Un savant Chrétien, J.B. Biot," *Le Correspondant*, Nouvelle Série 36 (1837): 955–995.

[12] For biographical information, see Jean Poujoulat, *Le Reverend Père de Ravignan; sa vie, ses œuvres* (Paris: Regis Ruffet, 1862); Emile Picard, *Les conférences Catholiques au XIXe siècle: Lacordaire et Ravignan: thèse* (Toulouse: A. Chauvin et Fils, 1871); Eugène de Mirecourt, *Ravignan* (Paris: Chez l'auteur, 1858).

[13] Quoted in Poujoulat (Paris: Regis Ruffet, 1862), 158.

[14] Ibid., 175.

[15] Ravignan, "Les droits de la raison (1838)," in *Conférences du Révérend Père de Ravignan de la Compagnie de Jésus* (Paris: Librairie de Mme Ve Poussielgue-Rusand, 1860), 391.

[16] Ibid., 392.

[17] Ibid.

Truth, then, lay in the realm of inaccessible light. Although the reasoned deductions of science were bound by the visible, Ravignan looked beyond to what lay outside the limits of sight. Biot's work as well, from the 1820s, began to center on what could not be seen. Polarization, as we shall see in later chapters, became his own form of inaccessible light leading him further away from the domain of the visible.

The zodiac of Denderah

Only a few weeks after his crushing showdown with Fresnel, Biot returned to stand before the Academy of Sciences again. "If in any case one circumstance can allow me to find grace before the Academy," he told them, "it is the particular nature of what I dare here to submit to here."[18] The subject of Biot's daring memoir was a highly controversial zodiac taken from the temple of Denderah in Upper Egypt. It depicted the heavens divided into twelve houses represented by the same symbols still used in astrological systems. This was the first time a zodiac had been found in Egypt and it promised a tantalizing glimpse into its origins. Most intriguing, the zodiac, if taken as a literal representation of the sky, could provide the first absolute date established for ancient Egypt, and would be invaluable for knowledge of the earliest moments of human history.[19]

Biot took aim at what he called the "dogma" of the immense antiquity of the zodiac that had infiltrated the Academy of Sciences.[20] The astronomers of Napoleon's expedition, he claimed, had set off for Egypt with the work of Charles Dupuis in their minds. Dupuis had published, shortly before, a text treating all religions as myth, and placing its origins in Egypt over 15,000 years ago.[21] When the expedition uncovered the zodiac in 1799, several of the members took up the project of working backwards to determine the point of time at which the heavens resembled the star arrangement depicted.[22] They came up with what Biot called "almost fabulous antiquity."[23] The date of the zodiac had clear religious implications. Standard Biblical chronology of the eighteenth century placed the flood around 2500 BC, so any evidence of human civilization would have to fit within that time period. Dupuis's dates had been

[18] Biot, "Mémoire sur le zodiaque circulaire de Denderah," reproduced in *Recherches sur plusieurs points de l'astronomie égyptienne appliquées aux monumens astronomiques trouvés en Égypte* (Paris: F. Didot, 1823).

[19] Jean Dhombres, "L'esprit de géométrie en Égypte. Monge et Fourier et Jomard: de la science conquérante à la science positivée," in *L'expédition d'Égypte, une entreprise des Lumières 1798–1801*, Patrice Bret, ed. (Paris: Éditions TEC & DOC, 1999), 327–349. Charles Coulston Gillispie, *Science and Polity in France: The Revolutionary and Napoleonic Years* (Princeton: Princeton University Press, 2004), 574.

[20] Biot, "Mémoire sur le zodiaque circulaire de Denderah," reproduced in *Recherches sur plusieurs points de l'astronomie égyptienne appliquées aux monumens astronomiques trouvés en Égypte* (Paris: F. Didot, 1823), xiii.

[21] Charles Dupuis, *Origine de tous les cultes, ou Religion universelle* (Paris: Chez H. Agasse, 1794).

[22] Jed Buchwald, "Egyptian Stars under Paris Skies," *Engineering and Science*, 4 (2003): 20–31.

[23] Biot, "Mémoire sur le zodiaque circulaire de Denderah," reproduced in *Recherches sur plusieurs points de l'astronomie égyptienne appliquées aux monumens astronomiques trouvés en Égypte* (Paris: F. Didot, 1823), xix.

part of an explicit attack on the foundations of religion. The astronomers of the Egypt expedition were scarcely more concerned with trying to conform to the Book of Moses.[24] Their dates, Chateaubriand suggested in his *Génie du Christianisme*, come largely from their "esprit d'irréligion."[25]

The debate was still raging in 1821, when the publisher Sébastien-Louis Saulnier cut the zodiac out of the temple in Denderah and brought it to France. The government bought it for 150,000 francs, and assigned Biot the task of recalculating the astronomical projection. He spent days on end carefully measuring every centimeter of the zodiac, benefiting from its extraction from the ceiling of the windowless dark temple. He assumed a more elaborate method of projection (i.e., the way that the Egyptians moved from the spherical surface of a globe to the flat surface of the monument), and corrected what he saw as the errors of his predecessors.[26] Taking all this into account, Biot then applied the formulas of the precession of the equinox and came up with his own date for the monument of 700 BC.[27] Having thus saved Biblical chronology, Biot sent a copy of his work to the king, hoping to curry favor (see Figure 3.1).[28]

A few weeks after he presented his work in 1822, Biot received a letter from a young philology student named Jean-François Champollion. The letter informed Biot that a few of the "stars" he had been using were not stars at all, but hieroglyphic symbols used to mark the ends of phrases. What Biot did not know, and what in fact no one knew at that point, was that Champollion was weeks away from cracking open the key to the ancient Egyptian symbolic language. He had been using a plaster cast model of the Rosetta Stone, which the Napoleonic expedition had been bringing back with them to France, before they were intercepted by Lord Nelson and the Stone wound up at the British Library. Champollion announced his success at deciphering the hieroglyphs in a work of September 27, 1822, which provided a dictionary and a grammar of the ancient language.

Champollion's greatest rival on the subject of hieroglyphs was none other than Thomas Young, the English physician who had sided with Fresnel on the wave theory of light. Arago acted here, as he had for the optical question, as one of his biggest supporters in France. But despite the support on hieroglyphs, Arago held himself at a certain distance from the mania surrounding the zodiac. "I knew how to escape the influence of a subject for which the obscurity has become proverbial!" was his statement on the matter.[29] He dismissed it with the suspicion that it was probably of Greek origin and not original Egyptian astronomy. Champollion's own work seemed to confirm the

---

[24] Buchwald, "Egyptian Stars under Paris Skies," 25.

[25] François-René Chateaubriand, *Oeuvres Complètes* (Paris: Firmin Oidot, 1843) 1: 253.

[26] Ibid., 30.

[27] Biot, *Recherches sur plusieurs points de l'astronomie égyptienne* (Paris: Firmin Didot, 1823), 118.

[28] Blacas, Chambre du Roi à Biot, 8 Septembre 1823, MS 4895, Bibliothèque de l'Institut.

[29] Arago, "Thomas Young, Biographie lue en séance de l'Académie des Sciences, le 26 Novembre, 1832," *Oeuvres Complètes* (Paris: Firmin Oidot, 1843) 1: 253.

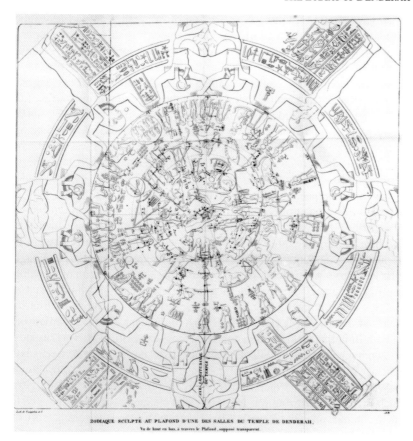

Fig. 3.1 Biot's drawing of the Denderah zodiac. (J.B. Biot, *Recherches sur plusieurs points de l'astronomie égyptienne appliquées aux monumens astronomiques trouvés en Égypte* (Paris: F. Didot, 1823).)

Greek origin theory, as he translated one of the hieroglyphs near the zodiac as autocrator, the Greek word for ruler.

But the zodiac question was thrown open again when Champollion visited Egypt for the first time in 1828. He returned to several of the sites of the Napoleonic expedition, copying down as many of the hieroglyphs as he could. His biggest surprise came in Denderah, when he could not find anywhere the infamous autocrator symbol that had relegated the monument to a post-Cleopatra chronology. Apparently, the cartouche in question had been blank, and an overly zealous artist had filled it in with a symbol seen elsewhere.

But another hint for Egyptian chronology appeared. In Saqqara, Champollion came across a series of inscriptions that seemed to explain the construction of the calendar and how it related to the agricultural year. The year was divided into three seasons of four months. Each month was 30 days long, which gave a total of 360 days. At the end of the year, 5 days of celebration were added to bring the total to 365. This number was still about a quarter of a day less than the true solar year, or how long it took for the earth to complete its orbit around the sun. The result was the année vague, or vague year, of the Egyptian calendar in which the beginning of the year was moved one day back every four years. Champollion admitted that the dates he was uncovering from the hieroglyphs remained impenetrable to him because he had not yet worked out the relationship between the vague year used by the ancient Egyptians and the true solar year. But he also found references to the two solstices and the vernal equinox, which follow the solar year, and he hoped to use these to tie the two calendars together.

Champollion needed help. When he returned to Paris in March 1830, he presented this problem to Biot. Armed with inscriptions referring to eclipses and solstices, Biot worked backward to establish the dates on which these must have taken place. The two men worked feverishly together. Champollion had also been appointed as the curator of the Egyptian wing of the Louvre, so he would typically spend his days overseeing the installation of an exhibit on all of the objects he brought back from Egypt, then he would spend his evenings at the home of Biot, providing translations for the items Biot had requested. It was during one such evening, on January 12, 1832, while working with Biot, that Champollion collapsed from exhaustion. He was rushed to his own home, but despite attentive medical care, he died several weeks later at the age of 42.

When Champollion died, much of his research was lost, and only that which Biot was borrowing was left. This, Biot claimed, was reason enough for him to continue on with his work.[30] But he also broadened the parameters a bit. Instead of merely trying to match up particular dates, he sought to establish the beginning date of the calendar itself. The Egyptian material offered one of the best glimpses so far of what Biot called, "the first ages of the world."[31] The key to doing this was what was known as the Sothic period. The Sothic period began when the first day of the Egyptian year happened at the same time as the helical rising of the star Sirius, which the Egyptians called Sothis. Although the two events started out on the same day, the first day of the Egyptian year was going to follow the année vague, and thus a year later would be a quarter of a day behind the helical rising, which followed the true solar year. As time passed, these two days progressively diverged until finally the vague year looped around and caught up with the solar year. Since for each year the difference was a quarter of a day, it took 4 × 365, or 1460 years, for this loop to occur, a length of time known as the Sothic period. Working backwards, Biot determined that the first day of the year matched up with the helical rising of Sirius in 275, 1780, and 3285 BC. This last date Biot took as the beginning of the Egyptian calendar, allowing him, in his words, "to go back through it to

---

[30] Biot and Champollion had presented the work jointly to the two Academies, but the memoir was authored by Biot.

[31] Biot, "Recherches sur l'année vague des Égyptiens lues à l'Académie des Inscriptions le 30 mars et à l'Académie des Sciences le 4 avril 1831," *MAS*, 13 (1835): 547–707, 551.

these first ages, and to effectively witness the development of the first notions of number and of time."[32] There is also another significance to the date that Biot does not mention. By this time, the Jesuits had pushed the date of the flood back about 600 years, making Biot's figure fall roughly at the Biblical moment for the earliest historical civilizations.[33]

But beyond the specific question of the Egyptians was a broader question about the nature and progress of knowledge. In 1830, the philosopher and failed astronomer Auguste Comte began publishing his work, *Positive Philosophy*, laying out his theory. Knowledge, he claimed, passed through three stages: the religious, the metaphysical, and the positive. Astronomy had been the first science to develop, and had always led others in passing through various stages. In ancient Egypt, astronomy was the definition of a science in the religious stage. To explain why things in the sky moved the way they did, Egyptians invoked supernatural causes and identified celestial bodies with Gods. It was in classical Greece, according to Comte, that astronomy passed to the metaphysical stage. The causes invoked to explain that the movement of the heavens was no longer supernatural, but they were abstract and undetectable, such as Aristotle's final cause which made the ethereal spheres of the celestial realm turn in perfect uniform circles. It was only recently, claimed Comte, that astronomy progressed to its final, positive stage where it no longer invoked any sort of mysterious cause, but rather dealt only in empirically verifiable laws.

Arago had a similar view of the process of astronomy. He dismissed the Egyptian vague year as a ridiculous construction inspired by "superstitious motives" related to the sanctity of religious celebrations.[34] "I undoubtedly have no need to insist on the puerility of considerations of this nature, which have since been abandoned."[35] But for Biot, issues of sanctity were far from puerile. Rather than seeing knowledge as a steady progression upward, he looked back to a past where knowledge levels had been very high, but had been subsequently lost. Furthermore, it was not clear that stripping science of its religious trapping would necessarily be a path to more knowledge. The astronomy of Egypt, after all, was clearly bound up with the religion of the day, and perhaps that religion would be the key to gaining access to this lost knowledge.

Biography of Newton

Biot shared an interest in ancient chronology with one of his long-standing scientific heroes: Isaac Newton. Newton had devoted a substantial portion of his later years to the project of sacred and profane chronology. He produced a series of shorter works culminating in the posthumously published *Chronology of*

---

[32] Ibid., 552.
[33] Maria Susana Sequin, *Science et religion dans la pensée française du XVIIe siècles: le mythe du Déluge universel* (Paris: Honoré Champion, 2001).
[34] Arago, *Astronomie Populaire* (Paris: Gide et J. Baudry, 1854–1857), 4: 672.
[35] Ibid., 672.

Ancient Kingdoms.[36] Newton was committed to the idea of prisca sapientia, in which the ancients had deliberately obscured the deepest religious mysteries in order to guard them from the common mind unable to handle them.[37]

In 1822, the same year he published his first essay on the zodiac, Biot also published a biography of Newton that made up the entry for France's *Biographie Universelle.*[38] He continued to research the subject, even visiting the archives in Cambridge, and became known as Newton's primary biographer in France. He was the first to attribute a previously unidentified text to Newton (the preface to the second edition of Johannis Collins' *Commercium Epistolicum*).[39] Newton had always been Biot's scientific model. He remained loyal both to the content, as in Newton's theory of colors, and the method. But his biographical account produced something of a scandal.

In his essay, Biot proposed that Newton underwent a mental breakdown in 1692 from which he never fully recovered. The occasion was a fire in his study, occasioned by a candle overturned by Newton's little dog, Diamond. The loss of years' worth on notes and manuscripts left Newton distraught to the point that he never again returned to the creative work that had distinguished him previously. Conveniently, this allowed Biot to separate Newton's work in gravity and optics (the *Opticks* was first published in 1704 but largely written by 1692) from his later work on chronology.

The reaction from Britain was shock and anger. David Brewster, Biot's former confidante, responded with an article defending Newton's reputation. It could not be true, he wrote, that Newton had labored in mental derangement from 1692, because that was the moment when Newton wrote his famous letters to Dr. Bentley on the Existence of Deity, which outlined a design argument for the existence of God.[40] Brewster claimed that this had long been the position of atheists like Laplace, who wanted to discount Newton's theological work by claiming he went crazy late in life.

Biot responded by clarifying that he had never intended to call the religious element of Newton's work into question. Indeed he admitted that Newton's interest in religion well predated the year of the mental crisis.[41] What Biot objected to were some of the specific claims made in Newton's later chronologies, claims that seemed to be rooted in an irrational anti-Catholicism. Newton

[36] Sir Isaac Newton, *The Chronology of Ancient Kingdoms Amended* (London: J. Tonson, 1728).

[37] Betty Jo Teeter Dobbs, *The Foundations of Newton's Alchemy* (Cambridge: Cambridge University Press, 1983), 108.

[38] Biot, "Notice historique sur la vie et les ouvrages de Newton," From *Biographie universelle*, 31 (1822): 127–194.

[39] J.B. Biot and F. Lefort, eds., *Commercium epistolicum J. Collins et Aliorum de analysi promota, etc., réimprimée sur l'éd. originale de 1712 avec l'indication des variantes de l'éd. de 1722, complétée par une collection de pièces justificatives et de documents* (Paris : Mallet-Bachelier, 1856).

[40] David Brewster, "Sir Isaac Newton's Illness and Refutation of the Statement that He Labored under Mental Derangement," reproduced in *Principien einer Elektrodynamischen Theorie Der Materie*, Johann Carl Friedrich Zöllner, ed. (Leipzig: Verlag von Wilhelm Engelmann, 1876).

[41] Biot, "Vie de Newton par le Docteur David Brewster, extrait du *Journal der Savants*, avril, mai, et juin 1832," in *Mélanges scientifiques et littéraires* (Paris: Michel-Lévy frères, 1858) 1: 237–290.

had, Biot pointed out in the article for the *Biographie*, come up with a chronology counting down to the apocalypse that treated the Catholic Church as part of the Grand Apostasy that could only be overthrown with the Second Coming of Christ. In Newton's reading of the Prophecies of Daniel, he identified the two-horned beast as the Catholic Church.[42]

Although passionately sharing Newton's interest in the history of sacred and profane time, Biot disagreed on many of the details. Biot pinpointed much of Newton's error in his dismissal of pre-Greek civilizations. He was particularly scandalized by Newton's dating of Egypt. To establish his dates, Newton had identified the Egyptian pharaoh Sesostris, founder of the first Egyptian Empire, with the figure of Sesac mentioned in the Bible, who lived in the reign of Solomon's son (thus establishing the primacy of the Hebrews over the Egyptians). Furthermore, Biot complained, Newton confused Osiris with Hercules, Isis with Cybele, Orus with Apollo, and collapsed everything into an abbreviated chronology limited to the first millennium BC. Far from dismissing Newton's work on chronology as irrelevant, Biot simply thought it was incorrect. Indeed, Biot continued his study of Newton over the next several decades. In the 1850s, he admitted to the particular interest Query 31 of the *Opticks* held for him. This query, which addressed God's action at the original moment of creation, was precisely the sort of thing that Laplace tried to separate from Newton's earlier work. But Biot turned seriously to the themes of idolatry and corruption presented in the query's last line.

> And no doubt, if the Worship of false Gods had not blinded the Heathen, their moral Philosophy would have gone farther than to the four Cardinal Virtues; and instead of teaching the Transmigration of Souls, and to worship the Sun and Moon, and dead Heroes, they would have taught us to worship our true Author and Benefactor, as their Ancestors did under the Government of *Noah* and his Sons before they corrupted themselves.

Biot, of course, did not give much credence to the particularities of Newton's account of early kingdoms. But that did not mean the project should be abandoned. The last Query of the *Opticks* was, after all, a query, offered to posterity, with the hope that future research may shed some light on the murky subjects. It was a shame, lamented Biot, that such questions had been abandoned.[43]

> If one found today among us a mathematician who was both experimenter and philosopher, would it not be a worthy and meritous work to take up now these doubts, expressed by Newton 140 years ago, and to examine what the discoveries made since then, in chemistry, physics, and

[42] Maurizio Mamiani, "Newton on prophecy and the Apocalypse," *The Cambridge Companion to Newton*, I. Bernard Cohen and George E. Smith, eds. (Cambridge: Cambridge University Press, 2002), 394.

[43] Biot, *Correspondance du chevalier Isaac Newton et du professeur Cotes, avec des lettres de plusieurs autres personnages éminents; Extraits du Journal des savants (cahiers de Mars, Avril, Mai et Juin* 1853) (Paris: Imprimerie Nationale, 1852), 22.

astronomy, would provide of facts, to shed light on them, better define them, or resolve them?[44]

But who, Biot asked, would be so indifferent to the applause of the crowd to undertake so solitary a task?[45]

The answer to Biot's rhetorical question seems clear. He often presented himself as virtually the only figure in France equally adept at the mathematical, experimental, and philosophical sides of science. And the discoveries of chemistry, physics, and astronomy were most likely his own. As we shall see in Chapter 4, Biot's work with optical activity lent credence to the query's claim that in the beginning God had imbued matter with certain active principles to account for the continual change of generation, vegetation, and fermentation, as well as the sustained emission of light from the sun.

And few people proved themselves so indifferent to applause. Even as he continued his work on the zodiac, Biot found himself increasingly marginalized by mainstream opinion. In the 1830s, the rising field of philology began to dominate discussion of ancient Egyptian objects. And most of these scholars of ancient languages viewed Biot's astronomical work as irrelevant and downright eccentric.

Leading the attack was the Greek scholar Antoine Letronne. He had weighed in over the Denderah controversy with a philological critique of Biot's astronomical method.[46] He claimed it was a waste of time to try to analyze the zodiac *astronomically*, since it had clearly been designed to serve an astrological purpose. The layout of the stars, Letronne claimed, could be interpreted in multiple ways and there was no use trying to match it to a particular moment in the past. What was more, the zodiac itself, with its division into twelve houses linked to star signs, was neither Eastern nor ancient knowledge. It was invented by the Greeks, and then moved from the west to the east, probably coming to Egypt during the period of Roman occupation. Chronologically, the twelve-house division began appearing on Greek and Roman monuments at the same time that man began believing he could read the future in the stars. The zodiac was thus founded on a false science and worthless as real knowledge.

Biot disagreed. He continued to maintain that the Egyptian zodiac was much older than anything in classical Greece.[47] But more than that, he rejected Letronne's criticisms that the astrology and religion could be separated from the astronomy. First, he pointed out, for the Egyptians every aspect of life, religious, political, and personal had been bound up for centuries with the passage of the sun. It was all but impossible that this relation left no material trace in the monuments of that period. The zodiac, like the entire arrangement of the

---

[44] Ibid., 23.

[45] Ibid.

[46] Letronne, *Sur l'origine grecque des zodiaques prétendus Égyptiens* (Paris: impr. de H. Fournier, 1837).

[47] Biot, *Mémoire sur le zodiaque circulaire de Denderah* (Paris: Imprimerie Royale, 1844). Also extrait des "Mémoires de l'Académie des inscriptions et belles-lettres," t. xvi 2e partie.

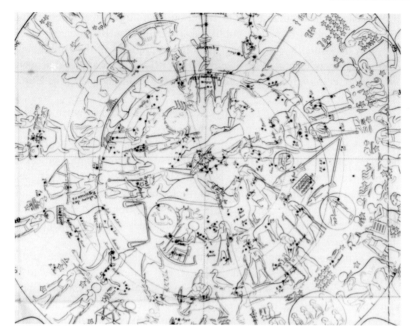

Fig. 3.2 Detail of Biot's drawing of the Denderah zodiac, showing the astrological figures. (J.B. Biot, *Recherches sur plusieurs points de l'astronomie égyptienne appliquées aux monumens astronomiques trouvés en Égypte* (Paris: F. Didot, 1823).)

temple, had been designed according to the marriage of religious ritual and astronomy (see Figure 3.2).

Chinese zodiac

Biot had also been interested in Chinese astronomy for a while. One of his first comments about the Denderah zodiac was that it reminded him of Chinese zodiacs.[48] His interest was fueled by the discovery of a large number of documents that had been previously lost or deliberately misplaced. The documents were reports brought back from China by the eighteenth-century Jesuit missionary Père Gaubil. Gaubil had been trained in astronomy before going to China, and had a particular interest in collecting information on Chinese astronomical practices. While in China, from 1723 until his death in 1759, Gaubil carefully examined every historical and astronomical chronology he could find. He then tried to match the chronologies to exact dates using eclipses mentioned in the records. His conclusion was that the actual chronology largely matched that provided by Chinese historians, and extended back to

[48] Biot, *Recherches sur plusieurs points de l'astronomie égyptienne* (1823), 5.

3000 BC. Gaubil had sent this text to his fellow Jesuit in France, Père Berthier, with the intention of having it published. Instead, Père Berthier quietly sat on it, then deposited it with other papers in the archives of the Bibliothèque Royale. It sat there forgotten for decades, until 1814, when Laplace discovered it while overseeing a transfer of records from the Bibliothèque to the Bureau des Longitudes.

It was not surprising that Gaubil's text caused some anxiety when it arrived in France in the middle of the eighteenth century. The Jesuit missionaries had left with the idea of incorporating the history of China into a biblical framework. The various nations of the earth would have arisen through a process of dispersal occurring sometime after Noah landed at Mount Ararat. This event, by the standard chronology of the time, took place in 2952 BC, and seemed hard to reconcile with the claim that Chinese history stretched back farther. By 1814, however, this issue was less tense because the Jesuits had pushed the date of the flood back about 600 years.[49]

Biot expressed excitement over Gaubil's chronology, but frustration with what had been left out. Gaubil, he claimed, had systematically excluded the astrological practices that accompanied the astronomical observations. Gaubil had tried to winnow the true science from the false, and thus extracted only what he needed while steering clear of pagan superstitions. Yet the astrological was, for Biot, the heart of the matter.

To recapture the original meaning of the practices that had been at once astronomical and religious, Biot needed another Champollion to make the original Chinese texts available to him. He found his translator in his son, Édouard. Biot had always had high hopes for his son, born in 1803. His first intention was to make an astronomer of him. He took Édouard on his geodesic expeditions to Dunkirk, and trained him in the use of the instruments. When he returned, he fixed up the little-used observatory attached to the College de France for his private use. He cleaned the telescope at his own expense, deposited his own instruments there, and had the College's architect renovate the building.[50] By the time Édouard was old enough to enter the Paris Observatory, however, Arago had consolidated his hold on the institution, and refused entrance to the son of his long-time rival.[51]

But by now Biot had another vision for his son's talents. Édouard returned to the College de France, where his father taught, to become one of the first students to enroll in the College's newly instituted courses on the Chinese language. After learning Chinese, he began translating Chinese astronomical texts for his father. His first major work was the first European translation of the text Tcheou-Pei.[52] This text, written during the Han Dynasty of the third century BC, described an astronomical system that stretched back millennia.

[49] Sequin, *Science et religion dans la pensée française du XVIIe siècles.*
[50] Biot, letter of August 9, 1819, Archives Nationales, F13 1083; "Vaudoyer, architect du College de France, à Bruyere," August 15, 1819, Archives Nationales, F13 1083.
[51] Picard, *Éloges et discours académiques.*
[52] Transliterated Zhou bi in Pinyin. Édouard Biot, "Traduction et examen d'un ancien ouvrage intitulé Tcheo-pei," *Journal Asiatique*, 3rd series 11, 593, supp. in 13 (1842): 198.

It contained a description of the Chinese zodiac, or the twenty-eight houses into which the heavens were divided.

Biot was particularly interested in the connection between the division of the heavens with politico-religious activity. His son's next task, accordingly, was translating a text bringing together precisely these elements. The Tcheou-li, or Rites of Tcheou, was a text that detailed the rites of the priestly class, and particularly the concordances of these practices with the phases of the moon. Tragically, Biot's son died before the work was published.[53] With most of the translation done, however, his father was determined to see the work through. Eventually, he put out an edition containing both the text and his own substantial analysis of Chinese astronomy.[54]

The key, Biot claimed, was that in the earliest known systems of astronomy, the study of the heavens was both science and theology. The division of the heavens, for both Egyptians and Chinese, was closely tied to the sequence of religious rites performed by the leader of the people. Egypt and China shared the identification of astronomy with a cult of the heavens, the assimilation of kings and the sun, and the use of figurative symbols in their language.

From this original state, a process of corruption followed: the Chinese had very advanced levels of astronomical knowledge, which they incorporated into their religious practices. This knowledge was limited to the priestly caste, however, and outside observers were only able to see the surface ritual. Part of Biot's insistence on the inseparability of science and religion was the conviction that the knowledge of the ancient East—although profound—was essentially occult, that is, marked by a deliberate impenetrability to the uninitiated. Astronomy, at its originating moment in Egypt and China, existed in the realm of the sacred. It was the movement from the sacred to the profane that led to error. The zodiac was the perfect example of this process of corruption. The moment when things went wrong was when the rites that had been intended to obscure genuine astronomical knowledge were interpreted by the uninitiated as the knowledge itself. The profane was elevated to the sacred, and genuine knowledge was replaced with idolatry.

But obviously he did not think of his modern scientific techniques as replacing the traditionally religious account of the origin of human civilizations. Rather, science became the principal tool for penetrating into previously veiled moments of the first ages of mankind.

Biot's exit from office, Arago's entrance

The Revolution of 1830 fundamentally transformed the grounds for political legitimacy in France, replacing the king's will with that of the people. It also ushered Biot out of power, and ushered Arago in. Biot's tenure as mayor

---

[53] Mémoire sur les travaux de M. Édouard Biot.

[54] Édouard Biot, *Le Tcheou-li ou Rites des Tcheou. 1–2*, par Jean-Baptiste Biot, ed. (Paris: Impr. nationale: B. Duprat, 1851).

of Nointel came to an abrupt end. The local inhabitants ransacked his estate in Nointel, and chased his family out. The oft-told story claims the peasants were inflamed by a portrait of Laplace, who they mistakenly took to be the hated king Charles X.[55] But perhaps the peasants of the Beauvais were less bumbling than history presents. Indeed it seems more likely that elegists and historians have been the only ones convinced by Biot's public claims of political disinterest, and that it was less Biot's portrait of Laplace than his deep commitment to traditional forms of authority that brought about his removal from public office.

For Arago, 1830 was a triumph. Siding with the revolutionaries, he had passed among the barricades himself and emerged as a major player in the political scene. On July 7, of that year, he won the election to become the permanent secretary of the Academy of Sciences. One of his first official duties was to deliver the eulogy of Fresnel, who had died on July 14, 1827. The day he was scheduled to deliver it was the same day that Charles X printed his revocation of the charter in the *Moniteur*, July 26, 1830. The Charter, signed by Louis XVIII in 1814, indicated that the king drew his legitimacy from a contractual relation with the people of France.

Arago's first plan was to scrap the speech, and instead deliver a short apology pointing to the "profound sadness" of recent events.[56] When he shared his plans with other members of the Academy, however, many of them protested and invoked the possibility of the suppression of the Academy. Arago relented. He read his original eulogy in its entirety, although certain phrases now took on an ominous subversiveness in light of recent events. Georges Cuvier, the perpetual secretary for the life sciences, demanded that he suppress the provocative phrases, but Arago insisted.[57]

Arago even managed to turn Fresnel's Bourbon sympathies into a condemnation of Charles X. When he mentioned Fresnel's support of the monarchy in 1814, he added, "The Charter of 1814, done without ulterior motive, appeared to him to contain all the seeds of a sensible liberty."[58] When Arago spoke of Fresnel's difficulty in finding a position, he ended with the phrase, that under such a poor administration, "a good citizen could fear that the future would not be free of storms."[59] In both instances, the audience met Arago's words with enthusiastic standing applause. Praise reverberated outside the Academy walls as well. Alexandre Dumas responded to the speech (which he labeled "a triumph") by claiming that Arago was a man of "not only science, but also conscience; not only genius, but integrity."[60]

---

[55] Picard, *Éloges et discours académiques* 261.

[56] Maurice Daumas, *La Jeunesse de la Science* (Paris: Belin, 1987), 155.

[57] Ibid., 156. For Cuvier's political conservatism, see Dorinda Outram, *Georges Cuvier: Vocation, Science, and Authority in Post-Revolutionary France* (Manchester: Manchester University Press, 1984); Toby Appel, *The Cuvier-Geoffroy Debate: French Biology in the Decades before Darwin* (New York: Oxford University Press, 1987).

[58] Arago, "Fresnel," in Arago, *Œuvres*, 3: 116.

[59] Ibid.

[60] François Sarda, *Les Arago: François et les autres* (Paris: Tallandier, 2002), 112.

When fighting broke out two days later, François Arago passed among the insurgents of the barricades of several neighborhoods. His brother, Étienne Arago, the director of a vaudeville theater, gained a certain notoriety by handing out weapons from the theater's stores. But François sought a more peaceful influence. In the afternoon, he sought an audience with Marshall Marmont, the military commander of Paris, charged with keeping the peace.[61] Arago did his best to persuade the Marshall to disobey his orders and cease firing upon the people. The king's actions, he argued, had broken the legitimacy of the crown. The people were only responding to violence with violence. The Marshall refused, and the Revolution continued to its bloody conclusion.

The Revolution of 1830 had swept critical public opinion into the halls of official power. The rational critical check on state authority became incorporated into the apparatus of the state, in the form of a legislative opposition. One of the most vigorous members of this opposition was Arago. In 1831, Perpignan elected him its representative in the Chamber of Deputies.[62] He seated himself on the benches of the extreme left and formed with a few other members a group that went under the label "radicals" (see Figure 3.3). For the next several years, he would, as the press put it, "signalize himself by an almost constant opposition to ministerial measures."[63]

France's first railroads

Arago's entrance into the Chamber of Deputies came at the right moment to highlight another disagreement between Arago and Biot, this time over the subject of railroads. Railroad construction had begun tentatively in the 1820s, and emerged as a controversial topic in the 1830s, with Arago and Biot on opposite sides. Their respective positions seem at first surprising. Biot, from his country estate, encouraged the production of the country's first steam-powered rail line, whereas Arago, the progressive reformer, did as much as any man in France to slow the country's rail development. Yet the puzzling reversal begins to make sense when viewed as an issue of public inclusion or exclusion. The disagreement was not for or against the technology per se, but over who would stand to profit from it.

Biot was a central backer of the first steam locomotive line in France. It ran from Saint-Étienne, site of France's most important coal mine, to Lyon, on the Rhone river. A canal had previously linked these strategic locations, courtesy of Louis XIV. After the Restoration, Biot found himself a substantial

---

[61] Daumas, *La Jeunesse de la Science*, 159.

[62] He had announced himself a candidate for both Paris, in the arrondissement of the Observatory, and the Pyrénées-Orientales. He won the election for the Pyrénées-Orientales on July 6, 1831. His election in Paris was forced to a second round, at which point he opted for the seat he had already won. Sarda, *Les Arago*, 121.

[63] *Sketches of Conspicuous Living Characters of France*, R.M. Walsh, trans. (Philadelphia: Lea & Blanchard, 1841), 283.

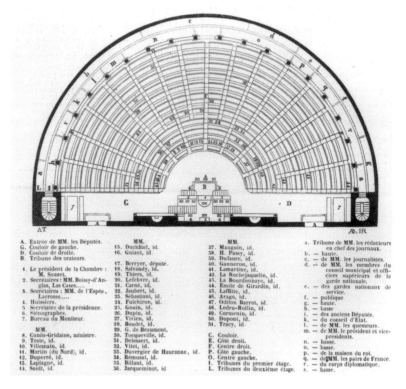

A. Entrée de MM. les Députés.
G. Couloir de gauche.
D. Couloir de droite.
B. Tribune des orateurs.

1. Le président de la Chambre : M. Sauzet.
2. Secrétaires : MM. Boissy-d'Anglas, Las Cases.....
3. Secrétaires : MM. de l'Espée, Lacrosse......

MM.
4. Huissiers.
5. Secrétaire de la présidence.
6. Sténographes.
7. Bureau du Moniteur.

MM.
8. Cunin-Gridaine, ministre.
9. Teste, id.
10. Villemain, id.
11. Martin (du Nord), id.
12. Duperré, id.
13. Lapiagne, id.
14. Soult, id.

15. Duchâtel, id.
16. Guizot, id.

17. Berryer, député.
18. Salvandy, id.
19. Thiers, id.
20. Lefebvre, id.
21. Carné, id.
22. Jaubert, id.
23. Sébastiani, id.
24. Fulchiron, id.
25. Gouin, id.
26. Dupin, id.
27. Vivien, id.
28. Boudet, id.
29. G. de Beaumont.
30. Tocqueville, id.
31. Delessert, id.
32. Vitet, id.
33. Duvergier de Hauranne, id.
34. Rémusat, id.
35. Billaut, id.
36. Jacqueminot, id

MM.
37. Mauguin, id.
38. H. Passy, id.
39. Dufaure, id.
40. Ganneron, id.
41. Lamartine, id.
42. La Rochejaquelin, id.
43. La Bourdonnaye, id.
44. Emile de Girardin, id.
45. Laffitte, id.
46. Arago, id.
47. Odilon Barrot, id.
48. Ledru-Rollin, id.
49. Cormenin, id.
50. Dupont, id.
51. Tracy, id.

C. Couloir.
E. Côté droit.
F. Centre droit.
P. Côté gauche.
O. Centre gauche.
I. Tribunes du premier étage.
L. Tribunes du deuxième étage.

a. Tribune de MM. les rédacteurs en chef des journaux.
b. — haute.
c. — de MM. les journalistes.
d. — de MM. les membres du conseil municipal et officiers supérieurs de la garde nationale.
e. — des gardes nationaux de service.
f. — publique.
g. — haute.
h. — basse
i. — des anciens Députés.
k. — du conseil d'Etat.
l. — de MM. les questeurs.
m. — de MM. le président et vice-présidents.
n. — basse.
o. — basse.
p. — de la maison du roi.
q. — de MM. les pairs de France.
r. — du corps diplomatique.
s. — basse.

Fig. 3.3 This is the seating chart of the Chamber of Deputies. The two front row seats in the left most banquette (marked as 45 and 46 here) were occupied by François Arago and fellow republican Jacques Laffitte. (From *L'Illustration* (1843), 132.)

shareholder and member of the canal's board.[64] Having recently returned from England, however, Biot felt it was his duty to inform the Board of the expansion of railroads he had witnessed. He gave a speech before them, warning that the only way to keep their canal from becoming obsolete was to undertake for themselves the building of a railroad to supplement it.

Biot's speech was met with reclamation of "Absurde! Absurde!"[65] But this only encouraged him to go out on his own. He became involved in the "Compagnie du Chemin de Fer de Saint-Etienne à Lyon." He recruited his brother-in-law, Barnabé Brisson, an inspector at the Ponts et Chausses, to put together the plans for the line after measuring the hills and valleys. The adjudication was granted in 1826 to a company headed by Édouard Biot and the

[64] Toussaint Cotelle, *Á M. le maire et au Conseil municipal de la ville de Beauvais, Mémoire ayant pour objet d'honorer le souvenir en cette ville du savant Biot* (Beauvais: Imprimerie C. Pere, 1879), 15.

[65] Cotelle, *Á M. le maire et au Conseil municipal de la ville de Beauvais*, 16.

Séguin brothers, Marc, Camille, Jules, Paul, and Charles, known for their work on steam engines for ships.[66] They completed the construction in 1832.

Édouard Biot continued work in the railroads, writing a manual on their construction in 1834. He noted that the Chamber of Deputies had turned its attention to the question, allotting funds to study the possibility of uniting France's major cities with railroads.[67] Arago, a recent member of the Chamber, took a leading role in these debates. Arago proclaimed himself as much a partisan of the railroads as anyone else, but only "on the condition that the mass of the public finds some profit in them."[68] He had some harsh words for the Saint-Étienne to Lyon line, which he accused of dropping its fare to wipe out the competition, then dramatically raising its fare as soon as it had established a monopoly. He proposed, in his first discourse on railroads before the Chamber, legislation to prevent this practice on future lines.

Arago really gained his reputation as a railroad skeptic in 1837, when the French government proposed a plan for a major rail system connecting several cities. Arago headed the eighteen-person commission at the Chamber of Deputies investigating the matter. His report on the subject sharply criticized the government's plans. He objected that the government would be too powerful. He took particular exception to "the spirit of monopoly which too obviously dominates the French administration."[69] The administration then withdrew its proposal, and many of the rail lines wound up being built by private companies. Rampant corruption ensued, and Arago later admitted that his opposition had been a mistake. He had been convinced that putting rail construction in the hands of the people would prevent abuses. Yet this would be one of the many ways the people would disappoint him.

Arago's public astronomy

If Arago's reputation was somewhat damaged by the railroad affair, he nonetheless received immense public acclaim from this work in astronomy. Although he was not named the Director of the Observatory until 1834, he had largely taken over the Observatory's workings by the 1820s.[70] His central goal was to bring transparency to the institution. He wanted the Observatory to be public, in every sense of the word. He devoted himself to efforts of astronomical popularization, and even incorporated the French public into the observing process.

[66] François Caron, *Histoire des chemins de fer en France, 1740–1883* (Paris: Fayard, 1997).

[67] Édouard Biot, *Manuel du constructeur des chemins de fer, ou Essai sur les principes généraux de l'art de construire les chemins de fer* (Paris: À la librarie encyclopédique de Roret, 1834).

[68] Arago, "Nécessité d'empêcher les compagnies de relever leurs tarif immédiatement après les avoir abaissés," *Oeuvres complètes*, 5: 233, extract from *Le Moniteur*, June 12, 1836.

[69] Arago, "Sur la nécessité de faire exécuter les chemins de fer par les compagnies," *Oeuvres complètes*, 5: 272.

[70] Sarda, *Les Arago*, 122.

Fig. 3.4 Arago delivering his lectures on popular astronomy to a packed hall at the Paris Observatory. (From *L'Illustration*, January 25, 1845.)

Arago's program of science popularization had two prongs: a series of free public lectures he gave at the Observatory and the publication of the *Annuaire*. Both had their theoretical origins in the Revolution, but none of them got going until Arago took up the job. The National Convention created the Bureau des Longitudes in 1795 and included as one of its duties the teaching of a public astronomy course.[71] The Bureau, however, only decided to begin honoring that particular statute on November 11, 1812.[72] They appointed Arago to give the lectures and allotted 1500 francs a year for the cause.[73] In 1841, the Observatory built a new hall to contain the crowds for the lectures. Arago himself wrote, "I am almost scandalised by the sumptuousness of my new amphitheatre" (see Figure 3.4).[74]

Arago's free astronomy course at the Observatory soon became a Paris institution. "All of Paris runs to hear them," an English paper wrote in 1840.[75] The courses stood as a public testament to Arago's "obstinate ardor in the cultivation of what the learned in x and y call the subaltern interests of the country and humanity."[76] Victor Hugo and Auguste Comte spoke of going. George Sand wrote begging Louis Blanc to accompany her when her usual escort was

[71] Arago, *Astronomie populaire*, 1: xi.
[72] Procès-Verbaux of the Bureau des Longitudes, November 11, 1812, *Bibliothèque de l'Observatoire*, 87.
[73] Archives Nationales, F17 3704.
[74] Lettter from Arago to Alexander von Humboldt, quoted in Karl Bruhns, ed., *Life of Alexander von Humboldt*, Jane and Caroline Lassell, trans. (London: Longmans, Green, and Co., 1873), 30.
[75] *Sketches*, 281. The preface describes that the sketches are translations of a series appearing weekly in Paris newspapers in 1840 and signed, "*un homme de rien*."
[76] Ibid., 273.

indisposed.[77] Even major names in science such as Jean-Baptiste Dumas, Élie de Beaumont, Boussingault, and Milne-Edwards attended.[78] In Hugo's *Les misérables*, one of the first things we learn about Combeferre, the most philosophical of the novel's young revolutionaries, is that he loved to attend public science lectures, where he "learned from Arago the polarization of light."[79]

In addition to the lectures, Arago also effected his campaign of public science through the Bureau's yearly publication, *L'Annuaire*. The Bureau originally intended this publication to regulate all the other almanacs. But when Arago was put in charge he also began including general interest articles, and its readership expanded to include a wide swathe of the literate public. "The *Annuaire* of the Bureau of Longitudes," a commentator wrote, "is read throughout Europe, and the articles of Arago on lightning, steam and the most delicate questions of astronomy have given it immense vogue."[80] Stendhal counted himself as "an assiduous reader," and wrote in asking Arago about his article on the effects of moonlight.[81]

Arago's efforts at science education dovetailed with his political ambitions. He was at once the foremost advocate of a public understanding of science, and one of the principal members of France's small Republican movement. These projects were on his mind: the forms of correct thinking he taught through astronomy would serve as a template and basis for the critical reason that grounded the Oppositional stance. His work aimed at sweeping away the vestiges of ignorance and superstition that formed the last obstacle to a properly functioning constitutional government. Grounded on the principles of equality and transparency, Arago's vision extended the enlightenment dream of wielding light as a sword. In the 1830s, the existence of this public took on a critical importance.

Comets

Politics and nature seemed tightly linked in the years after 1830, as France itself seemed to suffer under some dark star. It was the year of the "great fear of 1832."[82] Agricultural and economic crisis became mixed with meteorological aberrations. Poor weather led to crop failures. Wheat prices soared. Grain riots were endemic. Protests broke out against the forestry administration. The stock

[77] George Sand, *Correspondance*, Georges Lubin, ed. (Paris: Garnier Frères, 1964–1995), 26: 37. The letter, which states "Je voudrais votre bras si votre bras est libre, pour me conduire jeudi prochain au cours de Monsieur Arago," has no heading. The assignation of Louis Blanc is by the editor, who also dates the letter to the end of 1844.

[78] Daumas, *La Jeunesse de la Science*, 232.

[79] Victor Hugo, *Les misérables* (Paris: Gallimard, 1995), 664.

[80] *Sketches*, 281.

[81] Stendhal, *Correspondance*, III (Paris: Gallimard, 1967–1968), 200–201.

[82] M. Lucas-Dubreton, *La grande peur de 1832: Le choléra et l'émeute* (Paris: Éditions de la N.R.F, 1932), 107.

market collapsed with a spectacular crash.[83] Earthquakes in Italy were paired with political revolution there. Mysterious fires broke out in Western France, and were never fully explained.[84] Sometimes farmers blamed possessed young girls. The papers in 1831 covered the story of *la fille Choleau*, a young girl blamed for starting at least two fires. She had admitted to the fires, and claimed that she had been bewitched and pushed to light them by an evil spirit. At the trial, however, she remembered neither the fires nor the confessions. These "bizarreries" were all the more likely, the council pointed out, as she was several months pregnant.[85]

By far the biggest natural disaster of 1832 was the outbreak of cholera. The epidemic claimed 18,000 people.[86] Both the right and left together made the link between the disease and the political experiment launched in 1830.[87] It almost seemed to target the liberal opposition. Casimir Périer, the prime minister and one of the original deputies to support the July Revolution, died in May. General Lamarque, an opposition leader hailed as the regime's greatest protector of the people, soon followed. "For the past month," wrote the papers on April 30, "[cholera] alone has governed."[88] Arago himself fell ill.[89]

It was in this environment that reports began circulating of the comet of 1832. Papers began warning of a comet whose scheduled return brought it dangerously close to the Earth. The comet in question had first been spotted in late 1826 by the astronomer Wilhelm Biela. Soon after, the director of the Marseille Observatory, Adolphe Gambart, calculated its trajectory and determined that it had a period of six years and nine months.[90] Although its trajectory could be calculated exactly using celestial mechanics, its presence came wrapped in a discourse of portentous destruction. A first pass at the calculations revealed that the comet would pass directly through the ecliptic of the Earth's orbit, and the alarm went out that it might run into the Earth. Even if there was no actual collision, the papers mentioned the possibility that the Earth could pass through the nebulous material that made up the tail of the comet.

---

[83] Jardin and Tudesq, 106.

[84] Paul Gonnet, "Esquisse de la crise économique en France de 1827 à 1832," *Revue d'histoire économique et sociale*, 33, 3 (1955): 249–292.

[85] *Le National*, January 11, 1831. A quick look at how Arago handled another case of mysterious fires reveals his recurring strategy to use astronomy to undeceive peasant superstition. In 1842, a justice of the peace in Montierender faced a similar problem when a number of grain stacks in his area burst into flames with no explanation. This outbreak coincided, he observed, with reports of fiery orbs falling from the sky. He concluded that these must be aerolithes or meteors. He wrote a letter to Arago suggesting his hypothesis, and providing several examples of its probable occurrence. Arago enthusiastically accepted this explanation. He read the judge's letter before the Academy of Sciences, and urged all magistrates to consider this possibility before launching their proverbial or literal witch-hunts. *L'Illustration*, 1 (1843): 254.

[86] M. Lucas-Dubreton, *La grande peur de 1832: Le choléra et l'émeute* (Paris: Éditions de la N.R.F, 1932).

[87] Catherine Kudlick, *Cholera in Post-revolutionary Paris: A Cultural History* (Berkeley: University of California Press, 1996), 44.

[88] "Chronique de la Quinzaine, 30 avril 1832," *Revue des deux mondes* 5 (1832): 373.

[89] Daumas, *La Jeunesse de la Science.*

[90] Arago named it after the (French) man who calculated its return. Many people chose to call it Biela's comet, after the man who first spotted it. Arago, *Astronomie populaire*, III: 297.

The satiric chansonnier and "poète du peuple" Pierre Jean de Béranger sang of the end of the world in *La comète de 1832*.

> God is sending a comet against us;
> from this great shock we will never escape.
> I already feel our planet collapsing;
> The compasses of the Observatory will be lost.[91]

Like most of his chansons, this one had a political edge. Béranger had been a persistent critic of the Restoration, and his 1832 lyrics suggest that he did not find Louis Philippe a marked improvement.

> Haven't we had enough of vulgar ambitions,
> of idiots adorned with pompous titles,
> of abuse, of error, of extortion, of war,
> of lackey-kings, of a people of lackeys?[92]

Béranger welcomed the coming of the comet, and its destruction of the Earth, as a salvation from the bad governance of the lackey-king, Louis Philippe. And he was not the only one disappointed in the regime. After the funeral of General Lamarque, the unrest began to turn violent, and it looked like Paris might have another revolution on its hands.

As the barricades began going up, Arago faced a moral dilemma. The scene bore echoes of the revolution of 1830 that Arago had supported. But this time around, Arago was himself a member of the government. He had presented himself as the voice of the people, but in this case could do little to keep them under control. On June 6, 1832, Arago went with the rest of the deputies of the opposition, roughly thirty in number, to the house of Jacques Laffitte to discuss what position they would take with respect to the insurgency. They elected three of their members, Laffitte, Odilon Barrot, and Arago, to go speak to the king about ways to end the violence.[93] After entering the Tuileries in secret, they spent an hour and a half with the king, where they performed the difficult task of representing the people while rejecting their actions.

---

[91] Dieu contre nous envoie une comète;
à ce grand choc nous n'échapperons pas.
Je sens déjà crouler notre planète;
l'observatoire y perdra ses compas.
Pierre Jean de Béranger, La fin du monde et le commencement de l'autre, ou la comète de 1832 (Paris: Impr. de Sétier, s.d.).

[92] N'est-on pas las d'ambitions vulgaires,
de sots parés de pompeux sobriquets,
d'abus, d'erreurs, de rapines, de guerres,
de laquais-rois, de peuple de laquais?

[93] Different "transcripts" exist of this interview. Arago provided one, which exists as a manuscript in his family papers. Arago, "Entretien avec le Roi, 6 juin 1832," Papiers de la famille Arago, MI 372, *Archives Nationales*. Étienne Cabet also provided a transcript that he based on the discussion of the three men after their interview. Étienne Cabet, *Faits préliminaires au procès devant la cour d'assises contre M. Cabet. 4ᵐᵉ partie: Conférence du 6 juin 1832, entre S.M. Louis-Philippe et MM.Laffitte, Odilon-Barot, Arago* (Paris: Rouanet, 1833).

The opposition delegates stressed above all the need to ground the monarchy in public opinion. They made it clear that they did not share the insurgents' demands to abolish the monarchy of Louis Philippe. Yet they used the rioters' anger to make the point that the monarchy needed to return to the constitutional principles established in 1830. On Arago's first occasion to speak, he recalled the role of the 1830 revolution in changing the legitimate foundations of the state.

> France has accepted all of the consequences of the Revolution.—*Nearly all* of the members of the opposition wanted a *monarchy*, but a *popular monarchy*.[94]

The interview did not go well. The King insisted that the people loved him and were only goaded to rioting by the opposition's criticisms. Arago was at one point yelling so loudly that the King asked him to keep his voice down.

Arago was disappointed with the whole affair. Two days later, he wrote to John Herschel, complaining that "my *damnable* functions as a Deputy imposed quite onerous duties upon me during the deplorable events for which Paris has been the theater!"[95] The events of 1832 pointed to perhaps the greatest danger facing a constitutional government: the possibility that public opinion would not be sane and reasonable.

In the wake of the 1832 riots, Arago undertook a campaign to debunk forms of peasant superstition. The central superstition in his sights was the belief that the heavens had some influence over events on Earth. The question of celestial influence could only be answered, Arago claimed, through painstaking study of the various effects of the light emanating from celestial bodies. Comets were the first objects of his attention.[96] He first broached the subject with a several hundred pages article that dominated the 1832 *Annuaire* of the Bureau des Longitudes. The daily papers, he claimed, were spreading fear with their announcements that the comet would imminently strike the earth and smash it to bits.[97] Arago felt compelled to counter these claims with "everything that science has uncovered" about the trajectory of the comet, "whose proximity, we are assured, will surely be so fatal to Earth and its inhabitants."[98]

It was true, Arago admitted, that the equations of celestial mechanics predicted that the comet would pass directly through the ecliptic of the Earth's orbit. Yet a more thorough investigation, he added, revealed that there was no cause for fear. The point of crossing turned out to be over $4\frac{2}{3}$ Earth radii away from the orbit of the Earth. This distance was certainly close enough to allow for the possibility that the Earth would at least pass through the nebulous portion of the

---

[94] The italicized words are those which Cabet "guaranteed" as ones actually uttered. Cabet, *Faits préliminaires*, 3.

[95] "Arago to John Herschel, 8 July 1832," Archives of the Royal Society, H.1.356.

[96] His notebooks of that year contain reflections on the probability of a number of commonly believed superstitions, Bibliothèque de l'Observatoire, BL MSS Z5(1).

[97] Arago, "The Comet: Scientific Notices of Comets in General and in Particular the Comet of 1832," *The New World*, 65 (March 1843): 1.

[98] Arago, "Les Comètes," *Astronomie populaire*, III: 292.

comet. But, pointed out Arago, the comet was scheduled to pass by this spot on October 29, while the Earth would not get there until November 30. Given the fact that the Earth traveled at a speed of 674,000 leagues a day, one could see that it would never be less than 20 million leagues from the comet.[99]

And thus the comet's nebulosity would not transmit a physical effect to the Earth. Was there any other way the comet might influence worldly events? Popular prejudice said there was, and cited a host of incidents where unusual harvests followed comet sightings. Arago also set out to show the error in this view. First, he asked what kind of influence a distant comet could have on the earth. It was possible that it exerted a sensible gravitational pull as it passed by. Such an effect would have very limited consequences on earth, though. It could perhaps alter the tides a bit, but it would hardly bring about the change in temperature needed to affect crops. One might then think of the various radiations given off by the comet, and the additional heat they might provide. But here Arago pointed to the negative results of his research on the calorific effects of moon rays, adding that the light from a comet was much weaker in intensity than that of the moon.

A few years later, the return of Halley's comet once again focused public attention on the sky. Historians of astronomy have assigned this comet a particular place in a story linking the rise of celestial mechanics to a decline in celestial portentousness. Halley's comet was, after all, the first to have its orbit worked out according to Newton's laws of mechanics and gravitation. Eighteenth-century geometers hailed its return in 1759, within days of its predicted arrival, as the most convincing proof available of the application of mechanical law to the heavens. This triumph, however, did not necessarily dampen the public's enthusiasm for speculating about comets. Simon Schaffer has pointed to the many ways in which the cometographers' authority as prognosticators actually increased after 1759.[100] Comets themselves frequently retained their reputations as harbingers of doom. In France, Jérôme Lalande provoked an outcry of alarm with his 1773 *Reflexions sur les comètes qui peuvent approcher de la Terre*, suggesting that cometary orbits suffered marked variations, and this made it impossible to rule out collisions.

On Halley's return in 1835, Arago rushed to satisfy the public's curiosity and calm its fears in both the *Annuaire* and Observatory lectures.[101] He denied the widespread rumor that the comet had a noticeable effect on the year's harvest. Farmers had pointed out that the months of October and November were particularly mild in 1835, just as Halley's comet was visible in the sky. As the comet disappeared in December, a bitter cold set in. They drew the implication

---

[99] Ibid., 295.

[100] Simon Schaffer, "Authorized Prophets: Comets and Astronomers after 1759," *Studies in Eighteenth-Century Culture*, 17 (1987): 45–74.

[101] Arago gave reports on the comet's progress and properties at nearly every séance of the Academy of Sciences from August through December, *CRAS*, 1 (1835): 40, 66, 87, 96, 129, 130, 235, 255, 256, 322; he also wrote an account in *L'Annuaire* (1836); reproduced in Arago, *Œuvres*, XI: 481.

Fig. 3.5 An ad for a pamphlet entitled "The Comet: Past, Present, Future." On the left is a comet carrying water buckets entitled "The Comet of the Flood." On the right is a comet pulling wine casks entitled "The Comet of 1811," presumably recalling the year's exceptional harvest. (From *L'Illustration* (1843).)

that the comet provided additional heat. And yet, Arago warned, this account was not complete. The comet was still quite close to the earth in December, even if the public could not see it. It would have just passed by its closest point to the sun and had therefore, according to comet theory, replenished its heat for the long voyage out. That would mean that the comet had somehow heated the earth when it itself was cool, and cooled the earth when it was hot. Furthermore, Arago presented the results of his polariscopic measurements to show that the light of comets was entirely reflected and thus originated in the sun.[102]

The particularly dramatic comet of 1843 excited even more speculation. This comet fell in the "unexpected" category, as no one had predicted its arrival. It was, moreover, a phenomenon of striking brilliance, unmistakable in the night sky and even visible in daylight. On top of everything, it coincided with a devastating flood in the center of France, and an even deadlier earthquake in the French Antilles. Tracts such as *The Comet: Past, Present and Future* appeared for sale, promising "revelations, opinions, predictions, whims."[103] This pamphlet featured a coarse-faced comet in the dress of a peasant girl, with unruly *chevelure* (the word for hair and a comet's tail was the same) streaming behind her. She held two buckets of water, presumably to dump upon central France just as the title "the comet of the flood" was printed below her (see Figure 3.5).

More expensive journals such as *L'Illustration* treated the comet very differently. They printed a star map of a small section of the sky that would allow its readers to track down the comet for themselves (see Figure 3.6). It chastised the population for interrogating its astronomers "with an eagerness that has not always been enlightened." But nonetheless praised the public's interest in elevating their minds with the contemplation of the great laws of nature.[104]

---

[102] Arago, *Leçons d'astronomie professées à l'Observatoire Royal, recueilliés par un des ses élèves*, 4th ed. (Paris: Chamerot, 1845), 352; *L'Annuaire* (1836).

[103] *L'Illustration*, 1 (1843).

[104] *L'Illustration*, "La comète," 1 (1843): 64.

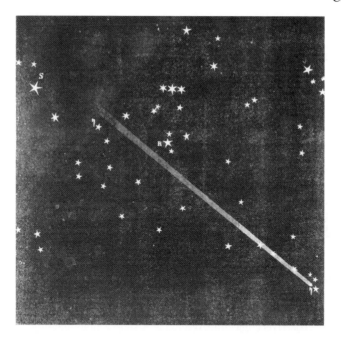

Fig. 3.6 A star map allowing readers to locate the comet of 1843. The attached description described Arago's observations, including his polarimetric measurements concluding that the comet shone with its own light, and not that reflected by the sun. (From *L'Illustration* (1843).)

The journal then presented Arago's report of the Observatory observations, including an account of the readings given by the polarimeter.

Arago also devoted an issue of *l'Annuaire* to the comet of 1843. "The common people (la vulgaire)," he pointed out, blamed it for everything from floods in the Midi to earthquakes in Guadeloupe. He assured his public that those responsible for "inventing or propagating" these complaints were "certainly strangers to the most elementary notions of science."[105] They were not (and this was their fatal flaw) able to "produce an argument either good or bad" for their claims.[106] Arago compared them to the Abyssinians, reported to tremble in terror at the great length of the comet's tail, or the Mexicans, who he claimed saw it as the portent of an imminent bonanza.[107] He told of his own research on the subject. A battery of meteorological data from around France gave no indication of anything unusual in the weather. He had also used the Observatory's most sensitive thermometric instruments to test the

[105] Arago, *Leçons*, 335, also in *L'Annuaire* (1844), Arago, *Astronomie populaire*.
[106] Ibid., 353.
[107] Ibid., 354.

light coming from various parts of the comet, and found that none of them produced a sensible calorific effect.[108]

Arago may have inspired his audience to think about comets, but that did not necessarily mean that he put an end to all their flights of fancy. One reader, Jules Verne, drew heavily from Arago's *Astronomie populaire* in the exposition of his 1877 novel *Hector Servadac*. But the plot line centered on just the kind of celestial disaster Arago was trying to downplay: a comet strikes the Earth, carrying off a bit of the Mediterranean and a few hapless passengers. One of the few buildings transplanted on to the comet was Arago's mountaintop shack on Formentera, where Verne's crackpot astronomer, Palmyrin Rosette, retired after being refused a place at the Observatory.[109]

Photometry

Arago dismissed astrology as superstition. "Astronomy has dissipated a thousand prejudices," he wrote, "It has overturned, it has reduced to nothing both judicial and natural astrology."[110] Arago compared the "absurd and shameful facts" in the annals of astrology to the "magnificent results" of astronomy.[111] A study of the heavens, based on conjecture, offered only suppositions and ridiculous evaluations in his eyes. Astronomy, he claimed, must be based on observation with the best possible instruments with which one can equip oneself.

One of Arago's proudest accomplishments was to add another instrument to the astronomer's collection. He used the principle of the polarimeter to develop a photometer capable of comparing the brightness of celestial bodies. Arago delivered his first treatise on photometry on August 5, 1833.[112] He described in it his idea for a new kind of photometer, which he continued to work on throughout the 1830s. In 1845, he presented an example of his instrument to the Academy of Sciences, along with the results of a large number of experiments.[113]

Photometry, Arago pointed out in his introduction, had long been one of the most important tasks of astronomy.[114] Measuring the brightness of a star

---

[108] Ibid., 353.

[109] Jules Verne, *Hector Servadac: Voyages et aventures à travers le monde solaire* (Paris: Hetzel Edition, 1877).

[110] Arago, *Astronomie Populaire*, 4: 774.

[111] Ibid., 775.

[112] "Séance de 5 aoust 1833," *Procès-Verbaux des Séances de l'Académie tenues depuis la fondation de l'Institut jusqu'au mois d'août 1835 Publiés Conformént à une décsion de l'Académie par M.M. les Secrétaires Perpétuels* (1916): 334. Arago reprinted the articles in his *Oeuvres complètes, Sur les moyens de résoudre la plupart des questions de photométrie que la découverte de la polarisation de la lumière a fait naître.*

[113] "Séance de 11 août," *CRAS*, 21 (1845).

[114] "Measurement" since Ptolemy typically involved simply looking at the star, and grouping it in a category with stars of similar brightness. This largely qualitative magnitude was then used primarily as a means of identifying and classifying stars. In the late eighteenth century, William Herschel and Laplace also began to treat magnitudes as indications of a star's distance from

was, along with determining its position, one of the few things that astronomers could do with these distant celestial bodies. The process was in principle quite simple. The human eye, he claimed, had no difficulty in accurately determining when two light sources had the same intensity. Yet difficulties arose in practice, as often in astronomy one wanted to compare the brightness of two objects that were not necessarily next to one another. One needed, Arago pointed out, methods of quantitative photometry to universalize the measurements in question.

The most notable efforts at quantitative photometry came in the eighteenth century, with Pierre Bouguer and Johann Lambert. In 1720, Bouguer introduced a method that involved visually comparing the light of a star to that of a candle, which he moved backward and forward until the two light intensities looked equivalent.[115] But these methods, Arago complained, were terribly imprecise. Even when two skilled observers such as Bouguer and Lambert measured the same thing (both men worked independently), they came up with very different results.[116] Arago did not attribute these differences to idiosyncrasies of perception. It was not difficult, he claimed, for any two observers to come to agreement about relative intensities. Rather, he suggested that the deviations may have arisen when the two light paths underwent varying degrees of reflection or refraction. This would have affected the state of polarization of the light, which in turn would have affected its intensity.[117]

Arago proposed to use the polarimeter as a more reliable means of generating a standard light source for comparison. The principle was largely the same: one would assign a number to the magnitude of a luminous body by comparing it to a standard, quantified source. His own photometer, however, would side step the flaws of previous light sources. Bypassing the light through two birefringent crystals, one could control its intensity in a consistent, predictable way. The first crystal would split the beam into two, with both emergent beams linearly polarized in opposite directions. One of these beams then passed through the analyzing crystal. As the intensity of the emergent beam varied according to the angle at which the crystal was rotated, one could use polarization to reduce the intensity of the light by a known amount, and have it serve as a standard for comparison.[118] Arago reported that he performed observations on celestial objects such as the moon with the first polarization apparatus he found, and it succeeded nicely.

Observations of this genre could be of immense public utility. For example, by comparing the light reflected directly from sun and light from earth, one could get an idea of how cloudy the earth was. Arago was most proud of his

---

earth. They then used these measurements to try to get a sense of the spatial arrangement of the universe. Arago, *Œuvres*.

[115] Pierre Bouguer, *Essai d'optique sur la gradation de la lumière* (Paris: Chez Claude Jombert, 1729).

[116] Arago, *Œuvres*, X: 151.

[117] Ibid., 152.

[118] Ibid., 261.

proposal to use the polarimeter to keep ships from running aground on rocks. As he pointed out, the sunlight reflected off the ocean would be polarized and the light from rocks would not. He could thus modify his device so that all the polarized light was refracted away. Anyone looking through it would then have an unobstructed view of the rocks ahead. He envisioned sailors the world over carrying around their polarimeters, averting danger on the high seas.

Polariscopic and photometric measurements lay at the heart of Arago's vision of a public astronomy. People across the globe would be able to make photometric measurements and send them in. Arago opposed a system of national observatories and instead envisioned a network of private citizens armed with the means of giving reliable data (see Figure 3.7).

This vision hinged upon the claim that the products of all observers' efforts were equal and interchangeable. To counter the evident fact that, using standard photometers, different people gave different results, Arago proposed an instrumental solution. The polarimeter, by regularizing and carefully controlling the polarization state of the light source, would alleviate the variation that plagued other devices.

(Papa, laisse-moi regarder ! — Tais-toi, je vois le noyau ! Enfoncé l'Observatoire !

Fig. 3.7 An example of public participation in astronomy. The child begs, "Papa, let me look." The father, looking at a kite through his telescope, responds, "Quiet, I see the nucleus! We beat the Observatory!" (From *L'Illustration* (1843), 77.)

In the 1830s, Arago created several new positions of "astronomes élèves," or assistants, to undertake the Observatory's new series of regular observations.[119] Although aware of the possibility of individual differences in their observations, Arago was convinced that these differences could be resolved with the right instrumental practices.[120] A good deal of Arago's scientific work in the 1830s involved getting to the point of treating everyone's observations as equivalent, by using the polarimeter to allow the easy comparison of far-flung observations.

For Arago, the polarimeter was the means by which a diverse group was to come to agreement with one another. The rationally communicating astronomical community he envisioned was predicated on the possibility that people could come to agreement with one another. But for this to occur they had to first agree on what it was they were seeing. Yet by the early nineteenth century, it was a well-known and increasingly documented fact that different astronomical observers saw things differently. Arago, concerned particularly with the problem of photometry, proposed that the problem could be remedied by using proper instruments. Two different observers might report on the intensity of single light source differently not because of differences at the level of observation, but because the light might, unbeknownst to them, have undergone a change in polarization that would affect its intensity. With a few modifications, Arago transformed the polarimeter into a device that could remove this source of error and render the photometric readings universal and interchangeable.

Conclusion

There is some irony in the use of the *Annuaire* as part of a campaign against astrology. The central purpose of the publication was to compile useful astronomical data for the coming year. Almanacs had frequently, in their history, had an astrological component. And many of the cheap almanacs that borrowed the calendar from the Bureau of Longitudes continued to offer astrological predictions.[121] The very project that gave the work its justification, describing what the skies would look like every night for the course of the year, seemed an impressive, and perhaps mysterious, act of prediction.

Arago, as the compiler of these predictions, appeared in the eyes of the public as the master of the skies. The power of the almanac was at the heart of one of the most famous scenes in Henri Murger's *Scènes de la vie de Bohème*, on

---

[119] M.M.L. Mathieu, *Histoire de l'Observatoire, Rapport fait au Bureau des Longitudes en réponse à la lettre de son excellence Monsieur le Ministre de l'Instruction Publique en date du 17 Mai 1864.* Archives Nationales F17 13569.

[120] Theresa Levitt, "I thought this might be at interest...: The Observatory as Public Enterprise," *The Heavens on Earth: Observatory Techniques in the Nineteenth Century*, David Aubin, Charlotte Bigg, Otto Sibum, eds., (Durham, NC: Duke University Press, 2009)

[121] John Grand-Cartéret, *Les Almanachs Français* (Paris: J. Alisié et Cie., 1896), LIII.

Fig. 3.8 Rodolphe burning his manuscript in Henri Murger's *Scènes de la vie de Bohème* (New York: The Century Company, 1924), 81.

which Verdi later based his opera.[122] The poet Rodolphe is in scene and is in his unheated Latin Quarter garret on a cold winter night. Rodolphe is inspired to write a poem to the woman he loves, but his hands are so cold that he cannot hold a pen. To gain some warmth, he begins burning the pages of the only thing he has, a play he had been working on for the past two years (see Figure 3.8). As the smoke curls out the chimney, the winds of Paris take notice. The winds hatch a plan to blow down the chimney and put the fire out. Just as they are getting underway, however, the south wind catches sight of François Arago, standing in the Observatory window and wagging a menacing finger at them.

At this point, the south wind cries out, "Let's get out of here quick. The almanac marks calm weather for tonight. We find ourselves in contradiction with the Observatory, and if we aren't home by midnight, M Arago will hold us to account."[123]

---

[122] Henri Murger's *Scènes de la vie de Bohème* (New York: The Century Co., 1924), 4.
[123] Ibid., 80.

So Arago makes an appearance as the master of the winds and the defender of the starving poets. The neighborhood housing Rodolphe's freezing garret was, by the way, in the jurisdiction Arago represented in the Chamber of Deputies. Arago's very fame made him appear in the popular press as the conduit between heavens and earth that he was denying.

The persistence of celestial influence was not for fiction alone. Biot mused upon how such a possibility would work. Light, he pointed out, was capable of a number of different effects. By considering the specific differences in action of different kinds of light, one was, according to Biot

> driven to consider generally the radiations sent off by material bodies, celestial or terrestrial, as composed essentially of an infinity of rays having diverse qualities and speeds, ... and which, according to their proper qualities can produce vision, heat, phosphorescence, determine certain chemical phenomena, and probably exercise beyond that, on inert material as on living bodies, many other actions which are still unknown to us.[124]

In the conclusion of his series of articles, he reiterated the importance of thinking of the world as inundated by a potentially infinite variety of different radiations, of which we could only be aware of the limited number we knew how to detect. Who knows, he asked, if the radiations from all ignatious bodies, either celestial or terrestrial, were the same? Perhaps each had specific properties that gave it a different influence over both inorganic and organic materials. Thus each different arrangement in space of these bodies would lead to effects that we ourselves would experience, either as new stars penetrated our planetary system, or as this system itself moved through diverse region of the universe. "Would one not say that science gives here some appearance of reality to the ancient prejudice that the stars influence our destinies!"[125]

The nature of light sat at the center of the question of celestial influence. Light, after all, was the vehicle by which information traveled from the heavens to Earth. But what was contained in this light? What impact did light have on human bodies? Was there some particular action between light and living organisms? Were there forms of light that had not yet been detected by human efforts? Arago and Biot both took up these questions. As we shall see in the following chapters, they did not always agree.

---

[124] Biot, "Sur les effets chimiques des radiations, et sur l'emploi qu'en a fait M. Daguerre, pour obtenir des images persistantes dans la chambre noire," *Journal des savants* (April 1839): 177.
[125] Ibid., 207.

4

A Vital Matter: Light and Life

Arago campaigned until his death against the notion of celestial influence. But after his death, it was a different story. At least according to J. Roze, a medium who regularly channeled Arago in his spiritualist séances of the 1850s and 1860s. In 1862, Roze published a book of his conversations with the Spirit of Arago.[1] Arago served, in death as he had in life, as a guide to the far reaches of the solar system. He described the topographies of the various planets, as well as the chief attributes of their inhabitants. Venusians, he let it be known, were intelligent but socially regressive, while Neptunians were moral but unsophisticated creatures. Jupitorians were the most advanced, having reached the stage where all beings existed in harmony and were able to levitate.

But the real point of astronomy, claimed the spirit Arago, was that it allowed one to think about the mystery of souls and the creation of life. "From here on," he claimed, "astronomy and theology are sisters and will walk hand in hand."[2] The question of souls had an astronomical resolution, each one developing out of what the spirit Arago called vital germs distributed throughout the universe. These germs, carried through the heavens by comets, had a particular law of attraction between them, which the spirit Arago compared to the attraction that a magnetizer exercised on a somnambule.[3]

There was clearly some irony in Roze's choice of Arago as his interlocutor. The same Arago who spent his time criticizing unfounded myths now peddled them. Arago the atheist now spoke of souls. There was one area, however, where the spirit Arago had it right: astronomical questions blended quickly into broader questions about the nature of life, matter, and spirit. The possibility of celestial influence hinged upon the interaction between the radiation emanating from the heavens and living things. Could the stars affect human destiny? It depended on what kind of light the stars were giving off, and what impact they had on the human body.

What was the relationship between light and life? Here, too, Arago and Biot diverged. Their own instrument, the polarimeter, rewrote the terms of the debate in the nineteenth century. Their work on polarization produced a

[1] J. Roze, *Révélations du Monde des Esprits* (Paris: Livre des Esprits, 1862).
[2] Ibid., 50.
[3] Ibid., 72.

phenomenon, optical activity, whose very name plunged it into the debate over active or passive matter. The claim that matter was active was a long-standing tenet of the vitalists, who insisted that life could only be explained by a vital principle. The materialists, on the other hand, held that matter was passive, and subject only to physical forces. The key to life was the ability of matter itself to take on a particular form of organization. Debates between vitalism and materialism flourished in the eighteenth century, generating much heat but little resolution.[4] Adding to the storm was materialism's distinctly political reputation, as it was blamed by many for the Revolution and the general crumbling of moral order.

Biot formulated the theory of optical activity while working at his country house in Nointel, defining it as the capacity of certain substances to rotate the plane of polarization of light. After decades of work, Biot determined that only organized bodies possessed this ability. His work offered, for the first time, evidence that there were physical forces which seemed to act only on living organisms. It became the biggest argument in favor of a renewed attack on materialism. Biot tried to push back the question of the mystery of life to its origins, in search of the divine spark animating the world.

Arago did not follow Biot in the discussion of optical activity. But while Biot was toiling alone in his country house, Arago occupied center stage in a drama unfolding in the Paris salons featuring the promise of living forces. A resurgence of Mesmerism swept through Paris, this time with Arago as the one to decide its fate. Arago surprised many of his colleagues by a tolerant interest in the occult, particularly in the claims of some people to "see" forms of invisible radiation. But his interest was essentially as a materialist. Far from invoking the supernatural, he sought to explain the strange phenomena as the physical effects of light. Indeed, "occult" may be something of a misnomer for Arago's interests. The literal meaning of the word refers to darkness and the unseen, but Arago focused on the somnambulists' promise to expand the domain of vision. Rather than citing some unseen cause as the explanation for their unusual abilities, Arago's somnambulists simply claimed a particular sensitivity to light. Although he repudiated animal magnetism by the end of his life, his dabbling can best be understood as part of a commitment to visibility, rather than invisibility.

Arago and the magnetizers

Light clearly had at least one major influence on living beings: it allowed them to see. The interaction between light and life was of crucial importance for the question of visibility. The act of seeing was, after all, the most obvious way in

[4] Peter Hanns Reill, *Vitalizing Nature in the Enlightenment* (Berkeley: University of California Press, 2005); Elizabeth Williams, *A Cultural History of Medical Vitalism in Enlightenment Montpellier* (Aldershot, Hampshire: Ashgate Publishing, 2003); Robert J. Richards, *The Romantic Conception of Life: Science and Philosophy in the Age of Goethe* (Chicago: University of Chicago Press, 2004).

which light had an influence on living beings. Light was the medium between the world and one's apprehension of it. But what were the limits of this process? How much of the world did light allow one to see? As a medium of transparency, it had to be admitted that light possessed certain drawbacks. There was so much, for example, that remained hidden. It was manifest to anyone with eyes that the world did not offer itself directly and equally. Rather, sight was a frail human thing, limited, and prone to failure. The reign of transparency and universality promised by the revolution seemed a long way off.

But just as the dream of transparency and universality seemed to wane in the political realm of the Restoration, a new form of universal vision became popular in the salons. In the 1820s, the dream of mankind breaking out of the limits of human sense surged once again. This time the question was the fad, growing in the 1820s, of seeing without the use of one's eyes. This fad was part of a resurgence of popularity of animal magnetism.[5] The notion that there were physical forces, otherwise undetectable, that could influence the bodies of living beings smacked of the mesmeric philosophy which had flourished in the 1780s. Franz Anton Mesmer had claimed that there existed a subtle fluid, analogous to the subtle fluids used to explain electricity, magnetism, light, and heat, which could be channeled and collected and used to treat various psychic ailments.[6] In the end, Mesmer was run out of Paris as a fraud, but not before establishing a fervent following among many of the city's wealthy and fashionable.

The magnetizers of the 1820s and 1830s reiterated Mesmer's claim that of a link between the physical effect and human will. François Noizet, a student of Arago's and one of the chief magnetizers, insisted that the human will played a role in the sensitivity of a person to light.[7] He accounted for the sensitivity in other body parts thus "I believe it is most simple to admit that the action of the soul is not fixed in a particular region, but it becomes sensitive everywhere the will directs it."[8] Demonstrations of the effect took place in salons throughout Paris. The Academician Léon Rostan, for example, placed a woman in a trance, then held a watch behind her head which she was nonetheless able to read perfectly.[9] André-Marie Ampère attended a session in which he played a game of cards with a magnetized man plunged in absolute darkness.[10]

[5] Nicole Edelman, *Histoire de la voyance et du paranormal: du XVIIIe siècle à nos jours* (Paris: Èditions du Seuil, 2006); Jacqueline Carroy, *Hypnose, suggestion et psychologie* (Paris: PUF, 1991).

[6] Frank Pattie, *Mesmer and Animal Magnetism: A Chapter in the History of Medicine* (Hamilton, NY: Edmonston Publishing, Inc., 1994); Robert Darnton, *Mesmerism and the End of the Enlightenment in France* (Cambridge, MA: Harvard University Press, 1968); Jessica Riskin, *Science in the Age of Sensibility: The Sentimental Empiricists of the French Enlightenment* (Chicago: University of Chicago Press, 2002).

[7] Bertrand Méheust, *Somnambulisme et Médiumnité (1784 1930), 1. Le défi du magnétisme animal* (Paris: Institut Synthélabo, 1999), 357.

[8] François-Joseph Noizet, *Mémoire sur le somnambulisme et le magnétisme animal adressé en 1820 à l'Académie royale de Berlin* (Paris: Plon, 1854), 107.

[9] Léon Rostan, "Magnétisme animal," *Dictionaire de médecine* , 23 (1825): 13.

[10] Jules Dupotet de Sennevoy, *Cours de magnétisme en sept leçons* (Paris: Roret, 1840), 435.

It was not, however, the mesmerism of the eighteenth century. Gone was the notion of an animal fluid. Just as physicists no longer spoke of light, heat, electricity, and magnetism as subtle fluids, so too did somnambulists switch their vocabulary to one of rays. Accordingly, the issue of light moved to the forefront, as the model form of radiation. The ether remained an important reference point, but it, too, was updated for the nineteenth century. Prior to 1820, the ether had been a fluid substrate through which light waves propagated longitudinally. Fresnel's work establishing the transverse nature of waves, however, forced physicists to reconsider the ether as a rigid solid, more crystalline than fluid.

In 1831, soon after being elected permanent secretary of the Academy of Sciences, Arago decided to revisit the question of mesmerism. The permanent secretary was a position of considerable power at the Academy, deciding which memoirs would be reviewed and what correspondence would be read. For the first time since Mesmer's disgrace in 1784, the Academy read a report on animal magnetism. Arago was adamant in his stance that claims about unknown forces should not be dismissed a priori, but should be evaluated the same as any other science.

In 1837, Arago found himself in the middle of a mesmeric controversy known as *l'Affaire Pigeaire*. The Academy of Medicine had offered a prize of 3000 francs to anyone who could read without using his or her eyes. Jules Pigeaire, a physician from Montpelier, stepped forward to claim it with the assertion that his daughter, Léonide, had the ability to read while blindfolded. Pigeaire arrived in Paris with his daughter in May 1838. Even before they met with the Academy commission, he turned their Paris apartment into an impromptu salon for demonstrating his daughter's abilities. They began conducting sessions on June 21, and held eleven of them over the course of the next several weeks. The author George Sand, the poet Théophile Gauthier, the editors of the *Charivari*, *Courrier Français*, and the *Journal du Commerce* attended. Of all the guests, however, Arago caused the biggest stir, as his scientific reputation lent weight to the proceedings.[11] The session he attended took place on July 7, 1838.[12] The girl sat, blindfolded, with a book placed behind a plate of glass before her. As Arago opened the book to random passages, the girl placed her hand on the glass and painstakingly made out the words.[13] The girl's father described Arago's participation as rapt:

> I could only with difficulty portray the astonishment painted on the fine face of Monsieur Arago when he saw the child read. He stayed speechless; his eyes, which rested on the somnambulist, had something particular about them which struck all the assistants.[14]

[11] Méheust, *Somnambulisme et Médiumnité*, 414.
[12] Ibid., 417.
[13] Bousquet, "Procès-verbaux," reproduced in Méheust, 597–603, 602.
[14] Jules Pigeaire, *Puissance de l'électricité animale ou, du magnétisme vital et de ses rapports avec la physique, la physiologie et la médecine* (Paris: Dentu and Germer Baillière, 1839), 96, quoted in Méheust, 417.

At the next session, the doctor Alfred Donné reported that Arago had been quite convinced and was searching for a theory to explain what he had seen.

Arago himself had little to say about the evening. He had even more reason to keep his mouth closed after the official commission had dismissed the girl's abilities as a fraud. In the end, Pigeaire and the commission could not come to agreement about the method used for covering Léonide's eyes. Pigeaire had been using a velvet ribbon, whose edges were stuck to the skin with a kind of tape. The commission, objecting that the ribbon was too thin to completely obscure vision, decided that they wanted to use their own covering. They first proposed a screen of black silk, to be placed 12 cm from the girl's face. The father objected that no light whatsoever could fall upon the eyes, or the phenomenon would not manifest. The commission then provided a black satin hood which covered a substantially larger portion of the face than the ribbon. The father again objected. It was essential, he claimed, that light from the book stimulate the facial nerves. Without this provision, not only Léonide would be unable to read, but also she would risk succumbing to dangerous convulsions. It would be endangering his daughter, he claimed, to allow the experiment to proceed with the hood. The commission thought otherwise, and the séance did not take place.

When the commission came out with its report, on July 28, 1838, they voiced their suspicions that Léonide Pigeaire had been cheating and refused to award the prize. The somnambulists, checked by their defeat, directed even greater pressure on Arago to come forward with a favorable report on what he had witnessed. Doctor Frapart, an advocate of animal magnetism, wrote letters, at first confidential but later published, goading Arago to speak. The academicians, he said, called the magnetizers fools.

> But if these facts are true, but if, having seen them, you deny them and dare not proclaim upon them; or if, having not seen them you do not want to verify them or bother producing them, then who, may I ask are you? Who are you? And what good do you serve?[15]

Frapart asked Arago, if not to sign the procès-verbaux, at least to attend another session, held in the salon of Carlotta Marliani, a friend a George Sand's. Arago remained silent. Frapart decided to use the silence to his advantage. He published a brochure with the following announcement:

> Even though this savant has not signed the procès-verbal, we affirm that he saw and saw clearly the phenomenon of vision through a thick velvet blindfold. If we are wrong, or rather if we are seeking to deceive, this affirmation that we deliberately make places on Arago the burden to immediately respond by a formal, positive, and public negation; but if what we say is true, this affirmation assures us that his silence, if it continues, is a favorable testimony.[16]

---

[15] N. Frapart, *Lettres sur le magnétisme et le somnambulisme à l'occasion de mademoiselle Pigeaire, à MM. Arago, Bazile, Bouillaud, Broussais et Donné* (Paris: Édouard Proux, 1840), 23; reproduced in Méheust, 439.

[16] Ibid., 25, reproduced in Méheust, 442.

Arago did keep his silence, which Frapart welcomed as an endorsement. The medical community, however, interpreted it otherwise. *La Gazette des médiecins practiciens* ran a notice stating that "M. Arago, having attested to nothing, let us advance that he has observed nothing and nothing was asked of him."[17]

But if Arago refused to pronounce on Léonide in particular, he did have some words to say about the controversy of seeing in the dark. When the physicist André-Marie Ampère died in 1839, Arago used the éloge as a venue for making a statement on the subject of occult activities. Ampère had come to dabble in occultism in his later years, and was particularly interested in the "exploring pendulum" that would set up oscillations in the presence of certain substances. This interest was generally not well received by his scientific colleagues. Michel Eugène Chevreul produced a particularly chastising critique of the activity, which he published in an open letter to Ampère in the *Revue des Deux Mondes.*

Entitled "On a particular class of muscular movements," Chevreul's letter explained that nothing very extraordinary was involved in the phenomenon of the exploring pendulum. Chevreul admitted that he himself had observed a pendulum begin to swing back and forth when he held it over water, mercury, or certain living beings. But, he pointed out, he had also observed several other things that led him to conclude that this motion was not the result of any mysterious forces. For example, the oscillations would diminish in relation to the arm being supported, and disappear entirely when the fingertips were supported. A person with their eyes closed would not be able to set a pendulum in motion, and if a person closed their eyes while a pendulum was moving, the motion would stop. Chevreul concluded that what was going on was that the pendulum was put into motion by some insensible movement of the arm. Once the oscillations began, the eye would follow them and endow the arm with a further tendency to motion. This tendency to move one's body in the direction of observed motion, such as when a pool player leans over while watching his ball roll, was both natural and unperceived. It was, he pointed out, why men, even those of good faith, could be deceived by this phenomenon and, in his words, "have recourse to completely chimerical ideas to explain phenomena that do not in reality come from the physical world that we know."[18]

Arago turned to this event in his official éloge of Ampère given at the Académie des Sciences. He admitted that Ampère had probably been taken in by frauds. But those who had so viciously attacked him were, he stated, just as unfounded in their excessive skepticism as Ampère had been in his excessive credulity. Ampère had been ridiculed because he thought that in a certain state of excitation a person could observe the stars with his knee. Well, said Arago, has it really been so well established that no man has ever, and no man will ever, been able to read in the absolute darkness of, say, the underground

---

[17] Ibid.

[18] Michel Eugène Chevreul, "On a particular class of muscular movements," *Revue des deux mondes* (1833): 158.

galleries of the Paris Observatory? Or has it been so well established that opaque screens, that is, screens impermeable to light, let absolutely nothing through which might produce vision?[19]

The answer, in 1839, could only have been "no" to both questions. The Academy of Sciences would have well remembered the near riots that occurred only months before when Arago had announced Daguerre's process for producing images from "photogenic rays." These rays could be separated by screens, yet, for Arago, still allowed one to see things. The same could be said for the calorific rays discovered by Melloni and investigated by Arago in the very underground galleries of the Observatory that he mentioned in his defense of Ampère.

The idea of an invisible chemical radiation was in fact fairly well established in Paris by the late 1830s. On the basis of the work of Macedonio Melloni, solar light was generally divided into at least three distinct groups: visible radiation affecting the retina, calorific radiation causing the sensation of heat, and chemical radiation instigating certain reactions. Melloni, an Italian working in Paris, conducted a series of experiments throughout the 1830s on the behavior of calorific rays.[20] He demonstrated that they could be entirely separated from the rays of visible light, and could be manipulated as distinct objects. He concluded from this that the two forms of radiation were physically distinct, and only happened to travel together on their path from the sun. Intrigued by the subject, Arago undertook his own experiments to establish the similarity of these forms of radiation with visible light, setting Hippolyte Fizeau and Léon Foucault on the task of getting interference pattern of calorific rays.[21]

Arago returned to these strange forms of radiation in a text "On Mesmerism," which he included in the 1844 éloge of Jean-Sylvain Bailly. Bailly had been one of the members of the Academy of Sciences' commission investigating Mesmer, and Arago praised his efforts to expose the charlatan. He described in detail the commission's tricks to catch examples of fraud. For example, the commission would magnetize one of several glasses of water, and note when the person claiming to be sensitive fell into convulsions after drinking from the wrong glass, or they would magnetize a certain tree in the garden and note when their subject was drawn to a different one. Their 1784 report was devastating, forcing Mesmer into exile and closing the case on his purported cures. But, Arago pointed out that this did not mean that one should dismiss animal magnetism out of hand. In some cases the apparently implausible may very well be amenable to reasoned explanation. Suppose, for example, that some people were able to see obscure forms of radiation, such as calorific rays. Then they would be able to see through various materials

[19] Arago, "Ampère," Notices biographiques, *Oeuvres de François Arago*, 2: 1–114, 89.

[20] Macedonio Melloni, *Mémoire sur la transmission libre de la chaleur rayonnante par différens corps solides et liquides, presenté à l'Académie des Sciences le 4 février 1833* (Paris: Bailly, 1833).

[21] Hippolyte Fizeau and Léon Foucault, "Recherches sur les interférences des rayons calorifiques (extrait)." *CRAS*, t. 25 (1847): 447.

usually considered opaque. This, claimed Arago, was perhaps the key to explaining several aspects of somnambulism that had up until then seemed implausible, such as people claiming to read through walls or read with their foot. So in the end, Arago's rather complicated support of the occult was not about unknowable, mysterious forces at all. It was about expanding the domain of visibility. Arago wanted to wipe out the last pockets of darkness in the world by discovering new ways to see.

The Baron Karl von Reichenbach struck a tone similar to Arago's. He criticized the "wretched magical trash" of Mesmer. What was needed was to "tear down the veil that hid his mysteries," and replace the "old phantasmagoria" with "sober scientific investigation."[22] Reichenbach claimed he was doing just that by proposing a new kind of radiation—the Od ray—that accounted for the variety of phenomena that went under the name animal magnetism. He first published his work in 1845 as a supplement in Liebig and Wöhler's *Annalen der Chimie*.[23] He had the support of Arago's friend Alexander von Humboldt, as well as a large French following.

Reichenbach was aware of the French work on polarization. Although, like Mesmer before him, he hailed from Southern Germany, he had traveled around France extensively in his youth, investigating the latest state of industrial works. He counted in particular Biot's report on the meteorite that had fallen on l'Aigle as one of his early inspirations. In the 1820s, he went into the beet-sugar industry.[24] Over the next few decades, the polarimeter came to play a larger and larger role within the sugar industry, using the power of its birefringent crystals to test the purity of sugar solutions. At the same time, crystals began to play a large role in Reichenbach's theory. Some of them, when placed in the hand of a sensitive person, provoked what he called a "tonic spasm." There was, he claimed, "in single crystals, a peculiar power, a fundamental, which had hitherto remained unobserved."[25] The force, he concluded, was polar, and had this in common with magnets. "The relation," he maintained, "between magnets and crystals to the animal nerve was entirely alike; while, on the other, the relation to iron, to the electric current, to magnetic poles, and to the magnetism of the earth, was, in magnets and in crystals, totally different."[26] Both magnets and crystals shared the property of attracting organic beings. Both of these were related to light. Sensitive people, in a darkened room, were able to see light surrounding a magnet, light surrounding crystals, and an "aura" of light surrounding living beings. Crystallization, he claimed, was the key to linking dead and living matter.[27]

---

[22] Karl Von Reichenbach, *Researches on Magnetism, Electricity, Heat, Light, Crystallization and Chemical Attraction in Their Relations to the Vital Force*, William Gregory, trans. (Seacaucus, NJ: University Books, 1974).

[23] Ibid., xxxiv.

[24] F.D. O'Byrne, "Introduction," In: *The Odic Force: Letter on Od and Magnetism*, Karl von Reichenbach, ed. (Seacaucus, NJ: University Books, 1968), xvii.

[25] Reichenbach, *Researches*, 35.

[26] Ibid., 46.

[27] Ibid., 63.

Reichenbach proposed that there was another kind of radiation, invisible to most people, that connected crystals, magnets, and the animal and vegetable kingdoms. He named the emanations "Od rays" after the Norse god Odin. They were, he claimed, "a cosmic force that radiates from star to star, and has the whole universe for its field, just like light and heat."[28] Od rays had poles to them, which Reichenbach described as "cool" and "warm," and which were also associated with complementary colors. A sensitive person would thus perceive one polarization as feeling cool and bathed in a blue aura, whereas the other polarization would feel warm and give off a yellow aura. Od rays could be polarized in the same way that normal light could, by reflecting them at an angle of 35°.[29] Reichenbach believed that up to one-third of the human population were able to see Od rays, and the rest were affected in ways in which they were not aware.

In 1845, a year after Reichenbach first published, the issue of light and magnetism was further tied together by the appearance of Michael Faraday's work on magnets and polarized light. He discovered that a very powerful magnetic field could rotate the plane of polarization of light, a phenomenon that became known as the "Faraday Effect." In July 1845, Faraday made an extended trip to France, visiting both Arago, who he described as "kind and pleasant," and Biot, who was "cheerful."[30] Faraday and Arago were well known to each other, ever since Faraday had explained the phenomenon, discovered by Arago and dubbed "the Arago effect," that a rapidly spinning copper disc could move a magnetized compass needle.[31] This experiment most likely turned Faraday's attention to inductive action.[32] Faraday's work in induction, where electrical phenomena could induce magnetic and vice versa, was couched in a philosophy that insisted that all physical effects must at some level be connected.[33] The claim, in 1845, that light and magnetism were connected extended this unity of forces.

The Faraday Effect made quite a splash at the Academy of Sciences.[34] Excitement about the connection between electricity and light prompted numerous French efforts to repeat and expand on Faraday's work. Alongside more sober attempts were the explicitly occultist Memoirs, such

---

[28] Karl Von Reichenbach, *The Odic Force: Letters on Od and Magnetism*. F.D. O'Byrne, trans. (Seacaucus, NJ: University Books, 1968), 23.

[29] Ibid., 22.

[30] Bence Jones, ed. *The Life and Letters of Faraday*, vol. 2 (London: Longmans, Green and Co., 1870), 217.

[31] In 1824, Arago had discovered that a rapidly spinning copper disc could move a compass needle. He had first noticed the effect while measuring the intensity of the Earth's magnetic field with Alexander von Humboldt. The needle of the compass continually came to rest faster when it was inside its box than outside of it. Arago then tried the opposite. If the needle was brought to rest by a still plate, what would happen with a moving plate? He found that the needle was displaced. Arago, "Du magnétisme de rotation," *Oeuvres*, 4: 424.

[32] Geoffrey Cantor, *Michael Faraday: Sandemanian and Scientist* (New York: Saint Martin's Press, 1991), 237.

[33] Ibid.

[34] First announcement by Dumas, *CRAS*, 22: 113.

as the one submitted by M. Ducros in 1847, entitled "Natural or Artificial Electrography," which claimed that certain words engraved into metallic surfaces and reflected by mirrors could be "seen" by those in a somnabulistic state, even at great distances and through opaque barriers rendering them completely invisible to most.[35]

Soon after, Arago found himself once again in the middle of another animal magnetism fad. In 1847, news reached Paris of an electric girl possessed of a particular sensitivity to the radiation of the world. Angélique Cottin was a fourteen-year-old girl living in a small village in the Orne Valley, where she worked at home making silk gloves. In 1846, she began manifesting strange effects. While she was spinning thread one day, the oak table on which she was working lurched violently. For the next several weeks all manner of furniture and objects were forcefully agitated in her presence. The villagers thought she was possessed or that evil spirits were behind the movement. The parish priest was called in to examine her, however, and he stated the opinion that demons were not involved, and that the strange phenomena had unknown but natural causes.

She was then taken to Paris, where she became "la jeune fille électrique." She was brought to the Observatory to be examined by Arago. She sat in a chair, which then began to move about violently. A table began to move when she touched it with a cloth. Arago also did some tests of his own. He had her hold her left hand up to a suspended piece of paper. He found that she did exercise a certain "repulsive action" upon the paper, but not drastically more than he had seen produced by other people. He also examined her carefully with a magnetic needle. He found no deviation of the needle, but nonetheless ended by recommending that the Académie des Sciences appoint a commission to investigate the matter further. The commission ended up deciding to dismiss the claims for Angélique Cottin's electrical nature. By that time, however, she was already something of a hit in the salons of Paris. And she became only one of an entire wave of electrical girls, the daughters of workers and peasants who would produce their effects in front of an audience of the educated bourgeoisie.

The legacy of mesmerism took another turn in the 1850s. The emphasis on invisible radiations was dropped in favor of the invocation of spirits. In 1848, another fourteen-year-old girl from a different rural village was once again causing furniture to fly about. The Fox sisters of Hydesville, New York, began hearing knocks in the walls of their house and soon began communicating with the spirit behind them, which they named "Mr. Splitfoot." The older sister, Margareth, was particularly sensitive, and the spirits advised her

---

[35] *CRAS*, 25 (Jul–Dec 1847): 29 "M. Ducros envoie un nouveau Mémoire ayant pour titre: Électrographie naturelle ou artificielle somnambulique avec lucidité, prouvée par la suspension de toute vision dans la fermeture des rebords des surfaces métaliques gravées irradiantes, et par les miroirs réfracteurs à plusieurs facettes, reproduisant, à des distances très-grandes et à travers de grands obstacles, au cerveau et aux yeux des somnambules, les images des mots gravés, sans l'existence, visible pour les autres hommes, de ces images dans ces miroirs."

to begin using a table to communicate with them. The sisters soon became famous and they started holding public séances across the United States. The phenomenon arrived in France in April 1853. The séance retained many of the same protocols in place in New York, such as seating arrangement and the translation of knocks into words. French speakers adopted the English word "medium," and even used the phrases "les mediums rapping" and "les mediums speaking."

We can see this transition in the work of Arago's friend, Victor Hugo. In 1843, he was inspired by recent developments in the study of light and photography to undertake a little dabbling in science himself. He came up with a theory of radiation that he articulated in three laws.

First law: The production of the so-called photogenic images without the aid of light, for example, in an cellar at night.

Second law: Magnetic vision.

Third law: With magnetic vision comes the as-yet unexplained phenomena of dreams, sympathy, ecstatic states, foreseeing, etc., an entire shadowy world that can be illuminated by this great law of radiation.[36]

The world of shadows would thus be illuminated by newly discovered forms of radiation that allowed unprecedented levels of sight. But this essentially materialist take on occultism would change after he went into exile for his participation in the revolution of 1848. In 1853, living on a desolate island of Guernsey, he began to undertake table-turning sessions with his family in which they communicated with departed or otherworldly spirits. These sessions would come to occupy several hours a day, every day, and for a number of years. The first spirit he contacted was his much mourned dead daughter, Léopoldine, but soon he moved on to writing poetry with Shakespeare, debating philosophy with Plato, and discussing matters of even greater import with Jesus and Death itself.[37]

Spinning tables and conversations with spirits were a far cry from the expanded forms of vision that had first caught Arago's attention.[38] The notion of universal visibility was gone, replaced by the action of unseen forces. This time, however, there would be no séance at the Observatory, and Arago would not be recommending that the Académie des Sciences appoint a commission to investigate it. The Academy of Sciences received its first report on turning tables soon after the phenomenon arrived in France in the 1850s. The memoir said that M. Séguin had witnessed a group of people seated around a table with only their fingertips barely touching the surface, and yet the table leapt

---

[36] This document of 1843 was first published in *Le Temps* on December 10, 1921. It is reprinted in Gustave Simon's introduction to *Chez Victor Hugo. Les tables tournantes de Jersey. Procès-verbaux des séances*, Gustave Simon, ed. (Paris: Editions Stock, 1980).

[37] Ibid.

[38] For a discussion of the turn to spiritualist explanations, see Bernadette Bensaude-Vincent and Christine Blondel, eds., *Des savants faces à l'occulte, 1870–1940* (Paris: La Découverte, 2002).

around as if subject to some great force. Because it was the work of a member of the academy, Arago acknowledged that it was his "duty" to present it. But he also voiced his own opinion at that session that the tables were of little scientific interest.[39] They were, he stated, the result of the vanishingly small impulsions given to the table by the fingers of the participants. Added up over time, these alone could be enough to move the table with considerable vigor. He cited a paper he recalled by the clockmaker Ellicott. Two pendulum clocks, in two different boxes, were hung on the same wall. Ellicott found that if he set one swinging back and forth, this motion would over time be transferred to the neighboring clock, implying that it had traveled by the imperceptible vibrations in the wall. This showed, Arago pointed out, that there already existed examples of movement being communicated in a manner analogous to that of the turning tables and thus their explanation "required none of the mysterious influences" usually evoked.[40]

Arago's final pronouncement on the subject came in 1853. Blind and on his deathbed, he oversaw the compilation of his collected works, including a section titled "animal electricity." He included here his negative reports on electric girls and table turning, and noted in his editorial remarks that no one had ever found a case of animal electricity that affected human beings. "One has vainly sought to associate it with the phenomena which take place in the human body, and on which human will only seems to have an effect for inattentive or biased people."[41] Human will, he pointed out, appeared independent from the influence of unseen forces.

Arago's involvement with occultism moved from entertaining the possibility of expanded vision, to rejecting the notion of spirits from beyond. The interaction of light and living beings always centered on the notion of visibility. He was trying to expand the domain of visibility to include the previously excluded. Arago became, for a brief moment, the somnambulists' best hope of legitimating the claims of seeing into the dark corners of the world. Their unusual abilities seemed to offer an expansion of light, all the while retaining its status as a purely physical force. As occultism veered into the realm of invisible spirits, however, Arago parted ways.

Biot's active matter

Biot, meanwhile, moved in the opposite direction. It was the idea of some spiritual, or at least nonmaterial, component of life that propelled him into an exhaustive study of polarization that would occupy the rest of his life. Not that Biot could be found holding hands around a turning table. His work was a far cry from the public spectacle surrounding animal magnetism. It was intensely

[39] Arago, "Phénomène des Tables Tournantes," *Oeuvres complètes de François Arago*, 4: 456.
[40] Ibid., 458.
[41] Arago, "Électricité Animale," *Oeuvres*, 4: 449.

personal, going unpublished for decades. Even then, it must have seemed to many an abstruse point in physical optics. But the implications were vast, for Biot had stumbled upon something of tremendous importance to the antimaterialists: evidence of a fundamental division between living and nonliving matter at the molecular level.

Biot's most valuable tool in tracing the boundaries of life was the polarimeter. He had noticed early on that certain organic substances, when dissolved in solution, could be used in the place of quartz crystals to polarize light, and that they gave much the same effect. Biot explained this as an effect of rotary or circular polarization: as the polarized light passed through the substance, its plane of polarization was rotated.[42] Their "optical activity," or the extent to which they rotated the plane of polarization, could be determined by the color of the extraordinary ray emerging from the polarimeter. The effect was intriguing: it appeared that living substances exerted some influence over the optical qualities of light that passed through them. But the number of active substances was still quite limited. By 1818, Biot had identified the property of optical activity in terebenthine, the essential oil of laurier, the essential oil of lemon, sugar syrup, and the dissolution of natural camphor in alcohol.

When Biot moved to Nointel, he began shining polarized light through every substance he could dissolve. He went to great lengths to establish the proper experimental conditions in his estate house: setting up a specially darkened room and outfitting it with all his polarization equipment. His experimental work also overlapped with his new-found agricultural interests. The living, growing products of his farm were perfect examples of the process of organization, by which an organism went from a relatively undifferentiated state, such as a seed, to a fully complex body. From his plants, he took substances such as sap that could be dissolved, and watched their progress throughout the season.

In 1830, he had a breakthrough. Previously, only a few substances had been strong enough to give a clearly visible signal. The technique he used to determine whether a substance was optically active was to place the dissolved substance in the polarimeter. The light would, of course, be colored. If the substance was optically active, the plane of polarization of the light rotated, and the color of the emergent beam cycled through the spectrum. To determine color with some precision, Biot sent the emergent beam through an achromatic doubly refracting prism. He always encountered a problem, however. If the solution sample was too thin, the planes of polarization of the transmitted beam would not be sufficiently dispersed. If the solution sample was too thick, too much of the transmitted light would be absorbed and the beam would be too weak. Also, the medium through which one observes the ray could itself be colored, which meant that it absorbed certain colors and allowed others

---

[42] This property is called "optical activity" in English and "pouvoir rotatoire" in French. For a more detailed study of how Biot's rotary polarization replaced Arago's chromatic polarization, see Jean Rosmorduc, *La polarisation rotatoire naturelle, de la structure de la lumière à celle des molécules* (Paris: Blanchard, 1983).

through. This, too, would weaken the coloration of the images. These difficulties, Biot reported, had kept him from discovering the full extent of rotary polarization.

These limitations disappeared, however, when he began using different features for his investigation. In particular, he found that, whatever the absolute quantity of rotation, there existed a direction of the rhomboidal prism, very close to the original plane of polarization, for which the extraordinary image given by this prism was an absolute minimum of intensity. If one assumed that the originally polarized ray contained all the elements of white light and that nothing was absorbed in the bending medium, then the proper tint of the extraordinary image in this minimum would be a violet purple almost exactly identical to the tint which, in Newton's construction, corresponded to the limit between red and violet. According to the proportions of light making it up, this purple should be a very good tint; nonetheless, it contained such a small fraction of the total transmitted light that, with very weak rotary action, it disappeared entirely. But immediately before disappearing, the extraordinary ray would take on a very distinct blue color; whereas immediately after the minimum, it would become a deep red. This set order of blue, nearly invisible purple, and red occurred within the space of 4° or 5°. As the ordinary image remained uncolored through the entire process, this became an extremely sensitive way of measuring the rotation of the plane of polarization. With his more sensitive device, Biot was able to discover optical activity where he previously believed it had not existed. He wrote

> Thus the sphere of these phenomena, by enlarging, has come to embrace a multitude of organic compounds and their numerous mutations, of which all of the phases have become detectable, appreciable, one can even say visible, by an index which is even more sure as it is inherent in the very molecular constitution of the bodies in which it is manifested.[43]

Virtually every organic compound he placed in his instrument now revealed this new and surprising property. What was more, Biot insisted that the property inhered in the matter itself and was not due, as was the inorganic crystal quartz, to the large-scale crystalline structure.

Biot claimed that he deduced this observation technique from the same principles he had demonstrated in the memoirs of 1812 and 1818. These works, justifying the use of Newton's tables to calculate polarimetric color, had been at the heart of the polemic between Arago and Biot. In his new memoir, Biot took up the same issue without ceding an inch, explicitly reiterating his analytical position on color in the first pages. The images that one observed through the polarimeter were, of course, colored. Biot maintained that these colors could be characterized numerically as a proportion of simple colors. Newton, Biot pointed out, had provided an experimental construction by which one could

[43] Biot, "Mémoire sur la polarisation circulaire et sur ses application à la chimie organique, Lu à l'Académie Royale des Sciences, le 5 novembre 1832," *MAS*, 13 (1835): 50.

calculate, for a given set of proportions, the particular color as it would appear to the eye.[44] Biot went farther than this. Newton had presented this work as the result of a number of experiments that he had conducted on the mixture of simple light. Since then, claimed Biot, it has been recognized that this result

> is linked by a very remarkable numerical relation to the lengths of the fits or light waves of the different simple rays; such that one can think that it is related to the physical nature of light, as well as its mode of action, much more intimately and deeply than Newton allowed us to see or per-haps than he himself was aware of.[45]

Here, as in the rest of his writings, Biot refused to make an issue of whether light was a particle or a wave. But, whichever the case, he insisted that Newton's color algorithms represented a physical reality.

The point Biot insisted upon, that light was compound and strictly phys-ical, was precisely the point that had animated the controversy of 1821. Far from losing to Arago and Fresnel and retiring to the country in admission of defeat, Biot regrouped his arguments and continued the attack. He recalled to mind the memoir of 1818, which had sat at the center of the debate over color. He had shown there the "minutieusement fidèle" agreement between the calculated and observed colors. He offered his current work as further evi-dence of his point. He had relied fundamentally on Newton's construction by applying his laws of rotation to the different rays that made up a simple color. He reported that he found this new verification matched "even more painstak-ingly than before." [46]

Biot's most comprehensive presentation of the previous decade's work came in 1832, with his, "Mémoire sur la polarisation circulaire et sur ses appli-cation à la chimie organique."[47] Its research agenda seemed straightforward enough: Biot proposed to use his polarization work to answer questions about the mechanics of chemical reactions. His optical techniques, he claimed, could detect molecular changes in instances where chemical analysis could not.[48] But this assertion was heavy with meaning, as it conjoined two more claims that Biot worked hard to establish: that optical activity was both a property of individual molecules and a necessary feature of organic material. This active nature was thus an intrinsic part of living matter that Biot could detect but chemists could not.

After meticulous, near obsessive work on the subject for almost two decades, Biot concluded that this optical activity was a property of all living matter. This distinction went far deeper than the chemical categories of organic

---

[44] Ibid., 42.

[45] Ibid.

[46] Ibid., 43.

[47] Biot, "Mémoire sur la polarisation circulaire et sur ses application à la chimie organique, Lu à l'Académie Royale des Sciences, le 5 novembre 1832," *MAS*, 13 (1835): 39–175.

[48] This aspect of Biot's work is well discussed in Seymour H. Mauskopf, "Crystals and Compounds: Molecular Structure and Composition in Nineteenth-century French Science," *Transactions of the American Philosophical Society*, 66 (1976): 5–81.

and inorganic. Organic chemicals that had been constructed in the laboratory from inert constituents possessed no such ability to rotate the plane of polarization.[49] Optical activity, Biot insisted, was not a property of organizational structure. It was, rather, the sole means in man's possession of confronting the otherwise undefinable limit between life and nonlife on the molecular level.

Biot had discovered early on that sugar gave particularly strong polarimetric readings. Much of his research, accordingly, had focused on this substance.[50] Traditionally, Biot pointed out, it had been defined by chemical means, specifically, as "those neutral substances which, dissolved in water and placed in contact with starch, decompose and transform into carbonic acid and alcohol."[51] Optical analysis seemed, however, to offer a means of distinguishing between sugars that appeared to be chemically identical.[52] Biot had studied a number of various sugars, either made by himself from vegetable materials or obtained from sugar producers, and found that they could all be identified by their particular optical activity.

A few months later, in January 1833, Biot teamed up with a younger chemist, Jean-François Persoz, to use optical activity to study the transformation of starch into sugar, the process that occurred during the maturation of fruits.[53] Chemists had been able to reproduce this process since 1811, by boiling the starch in a mixture of water and sulfuric acid, thus transforming it first into a gum-like substance, and then sugar. But Biot and Persoz were skeptical that the chemists knew what they were looking at. In his studies of the matter, Théodore de Saussure found that after he had boiled the water and filtered it, there was always a white coat left on the filter equal in weight to 4/100 of the original starch. He had assumed that this was part of the starch that escaped the action of the acid, and chemists had generally followed him in that assumption. But, said Biot and Persoz, Saussure was forgetting that starch had a heterogeneous composition, consisting of ovoid globules surrounded by a cortical envelope. The gummy material, they proposed, was in fact simply the globular interior denuded of its cortical envelope. Chemists were incorrect to say that the acid transformed the starch into gum, when all it did was remove the outer sheath. They brought in the techniques of rotary polarization to make their case. The globule exposed, they found, was not really a gum. It rotated the plane of polarization to the right, whereas natural gums rotated it to the left. For this reason, Biot named it "dextrine."[54]

Dextrin, the unexpected product of Biot's sugar research, would quickly become a lightning rod in the ongoing discussion of the border between life

[49] For example, "Biot à Pelouze, Novembre 1841," Archives of the College de France C.Xii Biot 22.

[50] Biot, "Mémoire sur la polarisation circulaire," 160.

[51] Ibid., 158.

[52] Ibid., 160.

[53] Biot and Persoz, "Mémoire sur les modifications que la fécule et la gomme subissent sous l'influence des acides," *MAS*, 13 (1835): 437–492.

[54] Biot, "Note Supplémentaire au mémoire précédent (Nointel, janvier 1834)," *MAS*, 13 (1835): 493–496.

and nonlife. It was, as Biot presented it, the only case of two bodies that, while chemically identical, could nonetheless be distinguished by the presence or absence of organization. As the chemist Dumas explained it, starch skated on the boundary between "organic substances," or inert chemical compounds and "organized substances," which formed tissues and organs, that varied in composition and belonged within the domain of physiologists.

> The characteristics of organization that starch takes on disappear indeed so quickly, so easily, and yet so completely, that we do not quite know how to go deeply into the study of this body. It seems to promise us some revelations on the relations by which the constitution of chemical compounds attaches itself to the much more hidden constitution of truly organized bodies, with which Chemistry refuses to seriously occupy itself at this moment.[55]

Dumas' comments came in the Academy report of a memoir by Anselme Payen, supporting Biot's work by showing the chemical identity of starch and dextrine. Dumas acknowledged that the question was important, and that Payen had answered it "as much as the state of science permits." His opinion, however, was that there was some as yet undiscovered way in which starch and dextrin were physically distinct.

Although Biot had not been a member of the reporting committee, he added his own opinion after reading the report. There was, he insisted, this physical difference between starch and dextrine: the one was *at that moment* organized and the other had lost this state of regular aggregation. When at the fecular stage, starch produced the effect on polarized light that only organized bodies did. When it had been desegregated into its components by, for example, the action of heat or acids, it no longer manifested these properties. On the basis of these observations, he claimed that it was the property of organization alone that distinguished the two substances, even as their nature stayed the same. Ultimately, the issue was that one possessed life and one did not possess life, and chemists claimed too much when they insisted that this difference had a material basis.

In 1835, Biot returned to Paris to continue his optical work. Although he maintained his property at Nointel for the rest of his life, and continued spending time there, he moved his polarization instruments to his laboratory at the Collège de France. Although he had been given access to the school's cabinet de physique, he decided to commandeer some extra space for his private researches. As he wrote to the administrator of the Collège de France,

> There are in the upper floors of the Collège de France two small rooms having each one a south-facing window, in which I have placed my polarization instruments: and I have set everything up around the benches to leave room for the products of my experiments and the various objects

[55] Dumas, "Rapport sur un Mémoire de M. Payen, relatif à l'analyse élémentaire de l'amidon et à celle de la dextrine," *CRAS*, 5 (1837): 898.

that I use. All of this is destined exclusively for my optical researches. I trust there will be no difficulty in this regard.[56]

Yet difficulty there was. The Director of Buildings responded to Biot's request with the statement that he was not allowed use of these rooms in the first place. The Collège architect had designated these rooms for the *Gardiens des travaux*, who were required to stay on the grounds of the Collège both day and night for surveillance. Biot, the Director pointed out, had ample space within his own apartments for personal possessions (as indeed Biot himself designated his polarization equipment).[57] Biot bristled at the "inspired pretension" of these demands and let the Collège know that, "it would be an unfortunate precedent to let architects meddle in our scientific arrangements."[58] After enlisting the support of the collected professoriate of the Collège de France, he eventually convinced the administration of his special needs.

Biot exercised tight control not only over the physical space of his work, but also over its material technologies as well. All of the polarimeters he used, he pointed out, were his own, and were made according to his own design. Biot used a white light source, but then placed red glass in front of the polarimeter to obtain a homogeneous light. The beam it produced was very weak, so that Biot did all his observing within a closed "cabinet obscure" with only the end of the polarimeter protruding.

In 1845, the optician Jean-Baptiste-François Soleil introduced a modification of Biot's design that increased the intensity of the beam to the extent that it could be used in the natural light of an open room. Soleil was hardly on friendly terms with Biot. He had worked closely with Fresnel and Arago, and constructed several demonstration instruments explicitly endorsing the wave theory of light.[59] When he billed his 1845 instrument as a "*polarimetre perfectionée*," Biot treated it as a personal attack. He denounced the modification as a "vice of construction" introduced by someone who had no understanding of what the instrument was really showing.[60]

Soleil's principal change was the addition of a quartz compensator, consisting of two plates of rock crystal, in the path of the light ray. The plates were aligned such that one plate rotated the light in one direction, the other in the opposite direction. The instrument's user began by turning the compensator

---

[56] "Biot to M. Sylvestre de Sacy, 15 October 1835," Archives of the Collège de France C.XII Biot 3.

[57] "Monsr. Guichard, le Directeur et President du conseil des batiments et monuments publics à M. de Sacy, administrator de College de France, 5 November 1835," Archives of the Collège de France C.XII Biot 5.

[58] "Biot à M. Sylvestre de Sacy, 24 avril, 1835, Nointel," Archives of the Collège de France C.XII Biot 11.

[59] G. L'E. Turner, "Soleil, Jean Baptiste François," *Dictionary of Scientific Biography*, Charles Coulston Gillispie, ed. (New York: Scribner, 1970) 519; G. Vapereau, "Soleil," *Dictionnaire Universel des Contemporains* (Paris: Hachette, 1861).

[60] Biot, *Instructions practiques sur l'observation et la mesure des proprietés optiques appellées rotatoires avec l'exposé succinct de leur application à la chimie médicale, scientifique, et industrielle* (Paris: Bachelier, 1845), iii.

so that the two halves of the split image were of equal colors. After placing the substance to be measured in the device, Biot wrote, one placed the center of one's pupil in the plane of junction of the two sheets of quartz, so that one received simultaneously the two images of the birefringent crystal, of different tints, one formed by the rotary actions in the "conspirant" direction, the other by those of the opposite direction. One would then turn the birefringent prism to the right or left until the tints were the same as at the beginning, and measure the angle traversed."[61]

But, claimed Biot, this technique involved a theoretical error. This double restitution of identity was only possible when the transmitted light was simple, in which case the direct measure of the deviation would be the same, and much more sure, without the intermediary of the two plates. When one used white light, which was the principal goal of this process, no position of the birefringent prism could give half-images that were rigorously of the same intensity and tint, although their differences may not be perceptible to the eye at low intensities.[62] Biot allowed that Soleil was very talented with his handiwork, and produced fine results as long as he followed Biot's directions. And Biot pointed out that he did not blame Soleil for the errors of his recent efforts.

> The abstract laws of the phenomena of deviations on which their measurement rests depend on a series of very subtle mathematical and physical considerations. Their study requires a set of theoretical antecedents that an artisan, even a very talented one, can never have the occasion to acquire.[63]

Soleil the instrument maker was not qualified to make decisions about what the polarimeter was depicting. He was unqualified for precisely the reason that Biot had laid out in the 1821 debate with Arago: his lack of analytical training made him unable to appreciate the complex nature of color. He literally did not know what he was seeing.

Biot's work after 1835 came to focus increasingly on the substance of tartaric acid. This acid particularly interested Biot for several reasons. One feature was that its optical activity appeared to change according to a very simple set of linear relationships as its concentration in solution changed. Biot sought to use this feature to address a long-standing question in chemistry: whether, when a complex organic product is dissolved in water, the atomic groupings remained distinct in a mixture, or whether they combined with one another to form new chemical combinations.[64] This was not always an easy question to answer, as efforts to remove a combination from solution (e.g. by the application of heat) could cause it to separate. Biot thus proposed his optical techniques to resolve the problem. He performed one experiment, in which he

[61] Biot, "Sur les moyens d'observation que l'on peut employer, pour la mesure des pouvoirs rotatoires," *CRAS*, 20 (1845): 1754.

[62] Ibid.

[63] Ibid., 40.

[64] Biot, "Question chimique proposée par M. Biot," *CRAS*, 1 (1835): 66–68.

poured boric acid into a solution of tartaric acid, and examined the effects of the resultant product on polarized light. He worked out the proportions in which the two substances combined, and offered his results as a challenge to the chemists. It would be interesting, he stated, if the chemists could come up with their own techniques to determine these proportions, so the numbers could be compared. But it would be even better if the chemists failed entirely:

> if, on the contrary, the invisible reactions that operate, with no apparent change, in diaphanous solutions are inaccessible to their research, perhaps they will find in this example a sufficient motivation to study closer the characteristics given by circular polarization, in order to weigh exactly the value of those indices they give on the actual molecular state of bodies; and in this case the goal I have set for myself will be completely achieved.[65]

The "goal" pervading Biot's work seemed to be some combination of promoting his own instrumental practices and forcing the acknowledgment that they allowed him to see things that others could not. He wanted chemists to rethink the indices they used to describe bodies in order to add optical activity to the list, and admit that it provided information about a body that chemical analysis never could.

Tartaric acid also seemed particularly useful for studying isomers. The phenomenon of isomers, that is, substances with the same chemical composition but different chemical properties, was first recognized in the 1820s.[66] As the issue became increasingly pointed in the 1830s, tartaric acid and its isomer, paratartaric or racemic acid, became the research sample of choice.[67] Biot established through polarization work that tartaric acid exhibited very energetic polarimeter readings, whereas racemic acid gave none at all.[68] This work provoked a response from the German chemist Eilhard Mitscherlich, who had spent the previous decade failing to find any discernible difference between a particular set of salts of the two substances, sodium-ammonium tartrate and sodium-ammonium paratartrate. By every technique of chemical analysis available, the isomers appeared to be not only chemically, but also physically identical (i.e. they both had the same atoms and these atoms were arranged in precisely the same way). The differences in optical activity, Mitscherlich concluded, must be accounted for by differences in crystalline arrangement, much in the way the optical activity of quartz was due to its crystal form.[69] This, of course, flew in the face of Biot's claim that the crystalline

---

[65] Biot, "Sur les combinaisons chimiques," *CRAS*, 1 (1835): 177–180.

[66] For a discussion of one of the more noted early cases, see J.H. Brooke, "Wöhler's Urea, and its Vital Force?—A Verdict From the Chemists," *Ambix*, 15 (1968): 108.

[67] Mauskopf, "Crystals and Compounds," 66.

[68] Biot, "Sur l'emploi de la lumière polarisée pour manifester les différences des combinaisons isomériques," *Annales de Chimie*, 69 (1838): 27.

[69] Mitscherlich, "Über die Gährung," *Gesammelte Schriften*, 537; see also Hans-Werner Schutt, *Eilhard Mitscherlich, Prince of Prussian Chemistry*, William E. Russwy, trans. (Washington, DC: American Chemical Society and the Chemical Heritage Foundation, 1997).

origin of the rotation of (inorganic) quartz was fundamentally distinct from the molecular origin of the rotation of organic bodies. In 1844, Biot presented the results that Mitscherlich had sent him to the Academy of Sciences. He took the opportunity, however, to assert that these chemical researches did not override his optical work.

> The only phenomenon whose observations and measurement could legitimately be related to the constituent molecular groups themselves consists *uniquely* in the deviation impressed upon the polarization planes of light rays, independently of their fortuitous state of aggregation, by a great number of substances, in truth, of organic origin.[70]

Biot insisted on the molecular origins of optical activity to preserve the hard distinction between quartz and organic matter. It did seem odd, Biot admitted, that two substances with identical molecular composition would give different polarimetric readings.[71] But he stood by his claim that the polarimeter provided information at the molecular level beyond anything the chemists could provide.

Biot's scorn for chemists was well known. He gained few friends among them when he referred to the lot as "nothing but a bunch of cooks."[72] Yet it would be a young chemist who solved his problem with tartaric acid. Louis Pasteur was working in a chemistry laboratory at the École Normale Supérieure in 1848. He had earned a doctorate there the year before and stayed to continue his research on the crystal forms of various tartrates. While working with sodium-ammonium paratartrate, he discovered that it actually consisted of two distinct crystalline forms. One form was the same as the sodium-ammonium tartrate and rotated the plane of polarization to the right. The other form was its mirror image, which rotated the plane of polarization to the left. When the two forms were combined in the sodium-ammonium paratartrate, their rotations canceled one another out. Here, then, was the answer to the problem: an identity of rotary power did indeed imply an identity at the molecular level.

Pasteur shared his results with Biot, who then asked to see them repeated.[73] Pasteur obliged and separated out the two forms of sodium-ammonium paratartrate. Biot himself then placed the left-handed variety into his polarimeter, and verified that its optical activity was just as Pasteur had predicted. "Then," Pasteur recounted, "the excited old man seized my hand and said: 'My dear child, I have all my life so loved this science that I can hear my heart beat for joy'."[74] This anecdote was a favorite of Pasteur's. His most noted telling of it took place in a lecture before the Société de Chimie in February 1860, precisely the moment when he entered into one of the most theologically and

---

[70] Biot, "Communication d'une note de M. Mitscherlich," *CRAS*, 19 (1844): 720.

[71] Ibid., 723.

[72] Archives of the Collège de France Cxii Biot 16.

[73] Pasteur, *Oeuvres de Pasteur*, Pasteur Vallery-Radot, ed., 7 vols., (Paris: Masson et Cie, 1922–1939), I: 325–326.

[74] Ibid., also quoted in Gerald L. Geison, *The Private Science of Louis Pasteur* (Princeton, CA: Princeton University Press, 1995), 57.

politically charged scientific controversies of that century in France: the debate over spontaneous generation.[75]

Félix Archimède Pouchet launched the debate in 1859 with the publication of *Hétérogenie, ou traité de la génération spontanée*.[76] In it, he presented his observations of the appearance of microorganisms in boiled hay infusions under mercury after the introduction of artificially produced air. Pasteur took up arms for the opposing side, and throughout 1860 presented several memoirs, indicating that Pouchet's results could be explained by contaminated air or mercury. He was utterly frank about the stakes involved in the debate. He declaimed to public audiences:

> What triumph for materialism if it could affirm that it rests on the established fact of matter organizing itself, taking on life of itself; matter which already has in it all known forces!...Ah! If we could add to it this other force which is called life....what would be more natural than to deify such matter?[77]

Materialism, with its relentless efforts to reduce everything to passive matter and mechanical forces, sought an explanation of the process of organization as the final step in a framework of fully physical explanations. Yet life resisted such an explanation, and remained irresolutely unmechanical, endowed on matter by God in what Pasteur called the "mystery" before one must bow.[78]

An interest in the force of life, and particularly in the way this force rotated the plane of polarized light that passed through it, formed the meat of the continuing interactions between Pasteur and Biot. In 1857, Pasteur wrote to Biot in Nointel telling him of his progress with his recent work on alcohol fermentation, and expressing the hope of using polarimetric readings to distinguish two products.

> You understand better than I do the interest there would be in establishing that the rotary phenomenon or its probable cause would distinguish it from the molecular arrangement, a phenomenon so general among the products of the organism, and related directly to the physiological role of these products, and that this quality of matter can have its direct interaction in the play of vital forces.[79]

---

[75] John Farley and Gerald Geison, "Science, politics, and spontaneous generation in 19th century France: The Pasteur-Pouchet debate," *Bulletin of the History of Medicine*, 48 (1974): 161–198; Bruno Latour, "Pasteur et Pouchet: Hétérogenèse de l'histoire des sciences," *Éléments d'histoire des sciences*, Michel Serres, ed. (Paris: Bordas, 1989); Antonio Gálvez, "The role of the French Academy of Sciences in the clarification of the issue of spontaneous generation in the mid-19th century," *Annals of Science*, 45 (1988): 345–365.

[76] Félix Pouchet, *Hétérogenie, ou traité de la génération spontanée* (Paris: Baillière, 1859); Jean-Louis Fischer, "Georges Pouchet (1833–1894): Le mouvement, la forme et la vie," *Le Muséum au premier siècle de son histoire*, Claude Blanckaert, Claudine Cohen, Pietro Corsi, and Jean-Louis Fischer, eds. (Paris: Editions du Muséum National d'Histoire Naturelle, 1997).

[77] Pasteur, *Oeuvres*, 2: 383–346; also quoted in Geison, *Private Science*, 111.

[78] Geison, *Private Science*, 111.

[79] "Pasteur à Biot, Lille 7 September 1857," no. 98, MS , Bibliothèque de l'Institut.

Pasteur hoped to use the "rotary phenomenon," or optical activity to probe the process of fermentation, which he was convinced was essential to the mystery of organization. Optical activity was found so extensively among living things, and was so directly related to their physiological role that it must be central in the mysterious process of life. The active "quality of matter" that it revealed interacted directly, Pasteur supposed, with the vital forces that rescued life from the realm of mere matter. Biot's discovery of an impregnable distinction between life and nonlife formed what Gerald Geison called the "glue" for the cluster of ideas weaving together scientific principles with a strong commitment to Catholicism and traditional France[80] Above all, it staunched the encroachment of a materialism that both men deplored.

Light played a very particular role for Biot. It was the strange nature of polarized light that revealed matter's subtle activity. Only by the most delicate of operations could Biot coax the minute change in color that signaled a rotation of the light's plane of polarization. The effect was recondite but undeniable: light was somehow interacting with whatever was alive in the matter.

In his own way, Arago, too, was after the elusive connection between light and life. Blindfolded readers and electric girls were unusual subjects for one of France's most eminent men of science. But Arago's interests were largely a continuation of his optical work detecting radiation. Mlle Léonide, sunk deep in her somnambulist trance, became another photometer. Arago used his new instruments to explore the boundaries of light's interaction with the human senses.

Biot and Pasteur made clear the enemy they sought to vanquish: materialism. But the materialists saw themselves as the heroes of their own drama, fighting against the dark forces of superstition. The vitalists saw the denial of the distinction between life and nonlife as the denial of whatever was transcendent in man. The materialists saw the intervention of spirit in the discussion of forces as a dangerous incursion into human autonomy. In both cases, the crucial implication was that of human freedom.

[80] Geison, *Private Science*, 135.

5

Light Paints Itself: The Conditions of Photographic Representation

～～～◦◦◦～～～

On August 19, 1839, the normally staid halls of the Paris Academy of Sciences were transformed into what one newspaper called a "genuine riot."[1] A crowd of unprecedented size had turned up to hear François Arago reveal Louis Jacques Mandé Daguerre's process for fixing the images of a camera obscura. The crush of people was already considerable at 10:00 am, four hours before the session was scheduled to start. By 3:00 pm, the crowd packed the hall and outer courtyard, spilled out on to the banks of the Seine, and "threatened disorder."[2] Utter silence prevailed as Arago went through the details of the process. Then the audience bolted for the doors, many of its members intent on trying it out while there was still daylight.

This frenzied scene stands in stark contrast to the tepid response that Daguerre had faced a year earlier. He had hoped to earn money off his process by selling subscriptions, but found few interested buyers. After months of fruitless trying, he approached Arago at the Observatory to ask whether he would lend his scientific name to an endorsement of the product. Arago, however, had other plans. He proposed that he would arrange for the French government to pay Daguerre a substantial pension if Daguerre would in turn disclose his secret openly to the public. Arago's efforts to convince the government of this plan took the better part of the next year, during which he criss-crossed the Academy of Sciences and Chamber of Deputies drumming up as much excitement as he could. The public announcement came on August 19, after the government agreed to pay Daguerre and the heirs of Niépce 10,000 francs. Daguerre had been slated to reveal the secret himself, but stepped aside because of "a little scratch in his throat."[3] Showman that he was, he ceded to the performer better suited for the venue.

---

[1] "Académie des Sciences," *Le National*, August 20, 1839.
[2] "Académie des Sciences," *Journal des débats*, August 20, 1839.
[3] This was Arago's phrase, Arago, "Le daguerreotype," *CRAS*, 9 (1839): 261.

The enthusiastic reception of August 19 seems a striking testament to Arago's skills as a publicist.[4] And nothing could be more true. Indeed, the fact that members of the public were there at all was evidence of Arago's deep commitment to publicity. Until a few years before, the only people who were allowed to attend sessions at the Academy of Sciences were members and those with special invitations. The general audience gained admittance for the first time in 1836, as part of the pitched battle over the public visibility of the Academy's proceedings. As permanent secretary of the Academy, and also as Director of the Observatory and Deputy of the Chamber, Arago had been actively involved in rendering the operations of these institutions transparent and comprehensible. The daguerreotype, as it traveled among these physical spaces, became itself a player in these debates over who should be allowed to participate in science.

Arago's chief critic in this business was his longtime disputant, Jean-Baptiste Biot. The two men had clashed before over the proper boundaries of the public in science. Biot, in his efforts to carve a position for the scientific notability modeled on the deposed nobility, advocated a science that remained behind closed doors. The photograph provided the opportunity for the old rivals to take up arms once again. The question they had argued about over the previous decades, "who had access to the secrets of nature?", became closely intertwined with the question, "what, exactly, did the photograph show?"

The epistemic status of the photograph, moreover, was deeply implicated within the material conditions of the respective practices of the two men. A great deal of work went into forging the links in the chain that connected the image to its referent, and this work can be found precisely in the site of labora-tory.[5] Arago and Biot ascribed two competing meanings to the photograph, which were accompanied by two strategies of investing it with reliability and two sets of experimental practice. Arago maintained the visible object as its referent. He invoked the mathematical precision with which the image was formed and relied on a medium that fully replicated this precision: the daguer-rian plate. Biot claimed that the sensitive surface in fact registered an invisible world inaccessible to the human eye. It was the unique chemical nature of the radiation that interested Biot, and he shunned the chemically suspicious silver plates for the more straightforward paper process.

---

[4] Talbot noted, with perhaps some bitterness, that "This great and sudden celebrity was due to two causes: first, to the beauty of the discovery itself: secondly, to the zeal and enthusiasm of Arago, whose eloquence, animated by private friendship, delighted in extolling the inventor of this new art, sometimes to the assembled science of the French Academy, at other times to the less scientific judgment, but not less eager patriotism, of the Chamber of Deputies," William Henry Fox Talbot, *The Pencil of Nature* (London: Longman, Brown, Green, & Longman, 1840); later historians have given Arago at least as much credit, see, for example, Helmut and Alison Gernsheim, *L.J.M. Daguerre: The History of the Diorama and the Daguerreotype*, reprint edition (New York: Dover, 1968).

[5] See, for example, Jennifer Tucker, *Science Illustrated: Photographic Evidence and Social Practice in England, 1870–1920*, PhD diss., Johns Hopkins University, 1996; Simon Schaffer, "Where Experiments End," in *Scientific Practice*, Jed Buchwald, ed. (Chicago: University of Chicago Press, 1995).

The mysteries within the sanctuary

It was only fitting that Arago occupied center stage in the announcement of the daguerreotype process. He was, after all, the man most responsible for making a stage of the Academy's hallowed halls. From its inception, the *Académie Royale des Sciences* had been an institution that drew its legitimacy from its social isolation.[6] By Arago's secretaryship, the Academy had grown sufficiently porous to inhabit the international Republic of Letters.[7] But, as Arago made clear in his work on electoral reform, the only true republic was a republic open to all. He thus threw his efforts into making public what went on behind the Academy's closed doors.

The Academy elected Arago as its permanent secretary for the physical sciences on June 7, 1830. By July 19, he was subverting its protocols enough to raise eyebrows internationally. "The volcano has come to an eruption," Goethe wrote from Berlin, "everything is in flames, and we have no longer a transaction with closed doors!"[8] The occasion was a memoir by Cuvier on the dodo of Mauritius, presented to the Academy on July 12. When Arago, as permanent secretary, read the procès-verbal of the July 12 meeting the following week, he included not just the title of the paper, as was usually done, but a short extract from it as well. The point was not a trivial one. The procès-verbaux were the only public records of what occurred within the Academy walls. Arago's extract lifted the rug of the smooth assurance of Cuvier's title to show the fractious debate underneath. Cuvier, the permanent secretary for the natural sciences, interrupted Arago before he was finished and complained that he was flouting the customs of the Academy. The issue eventually developed into a very public controversy between Cuvier and Geoffroy over the fixity of species.[9] Yet not the least of the consequences was, in Goethe's words, "that the debates of 1830 have modified the customs of the Academy and that was especially manifested in the meeting of 19 July."[10]

Arago also enlisted the popular press in his efforts to render visible the inner workings of the Academy. Declaring himself "an ardent friend of publicity and of the most entire freedom of discussion," he arranged for a special, heated room for the press where he made pertinent documents available to them after

---

[6] Mario Biagioli argues that the king's patronage required a "shield" to keep the Academicians' potentially uncivil behavior from reaching the public eye. "Etiquette, Interdependence, and Sociability in Seventeenth-Century Science," *Critical Inquiry*, 22 (1996): 193–238.

[7] Roger Hahn, *The Anatomy of a Scientific Institution: The Paris Academy of Sciences, 1666–1803* (Berkeley: University of California Press, 1971); Maurice Crosland, *Science Under Control: The French Academy of Sciences, 1795–1914* (New York: Cambridge University Press, 1992).

[8] Johann Wolfgang von Goethe, "Reflexions de Goethe sur les débats scientifiques de mars 1830 dans le sein de l'Académie des Sciences, publiées à Berlin dans les Annales de critique scientifiques." *Annales des sciences naturelles*, 22 (1831): 179–188; J.P. Eckermann, *Conversations of Goethe with Eckermann and Soret*, John Oxenford, trans. (London: George Bell and Sons, 1874), 479.

[9] Toby Appel, *The Cuvier-Geoffroy Debate: French Biology in the Decades Before Darwin* (Oxford: Oxford University Press, 1987), 159–161.

[10] Goethe.

the sessions.[11] Jokingly called "the sixth class of the Institut," this service was entirely the product of Arago's initiative, and had in fact never been discussed by the Academy.

He soon went further still, requiring the Academy itself to publish a full account of its proceedings in the biweekly *Comptes Rendus*. In addition to the flurry of publications, Arago sought to instantiate his project of transparency within the very architecture of Academy. Since his election as the perpetual secretary of the physical sciences section in 1830, he had, against much resistance, succeeded in creating a viewing area where the public could come watch the sessions.

Biot emerged as the most outspoken critic of the project. In an article published in the *Journal des savants* in 1837, he outlined the argument against the vulgarization of the Academy of Sciences. The presence of the public at academic debates would, he claimed, give a "different direction" to them than that required by the sciences.[12] If the Academy became a public venue, the savant would be forced to establish his reputation by impressing the audience. Doing so, however, he would have to seriously compromise the rigor of his presentation, as it was impossible to expect the majority of the audience to have a deep understanding of science. Science, for Biot, was a process best nurtured in the dark.

These debates resounded within the popular press. The official papers praised Biot as "a serious savant who does not look to popularize science by bringing it down and stripping it of the precision which gives it dignity."[13] They cast Arago, meanwhile, as the "Cromwell of the Observatory" who "vulgarized science."[14] The opposition papers leapt to the defense of their man with several front-page articles. They claimed the reaction to Arago's proposed changes was, well, reactionary. "Without doubt," *le National* stated, "if they lived in ancient Egypt they would have punished with death the priest hardy enough to divulge the mysteries of science closed within the sanctuary."[15]

The daguerreotype enters the picture

The popular press thus painted Arago as a divulger of mysteries, the man who made secrets public. In 1838, he would come into possession of one of the most mysterious secrets around: the ability to fix the images produced in a camera obscura. In that year, Daguerre paid him a visit at the Observatory,

---

[11] "Arago à Raspail, 21 mars 1835," MS 2388, *Bibliothèque du jardin des plantes* (also see AN MI 372).

[12] Biot, "Sur l'institution récente des comptes rendus hebdomadaires de l'Académie des sciences et sur la publicité donnée à ses séances," *Journal des savants*, 1837; reprinted in Biot, *Mélanges scientifiques et littéraires* (Paris: Michel Lévy Frères, 1858), 260.

[13] "Feuilleton de l'Académie des Sciences," *Journal des débats*, April 22, 1843.

[14] "Feuilleton de l'Académie des Sciences," *Journal des débats*, August 7, 1839.

[15] "François Arago," *Le National*, April 5, 1840.

Arago's personal principality of managed visibility.[16] Daguerre himself had long been in the business of marshaling light to the service of resemblance, having established his reputation by painting stage backdrops and inventing the diorama. He hoped, in a similar manner, to find profit in his pictures from light. He approached Arago, as well as several other notable men of science, including Biot, Alexander von Humboldt, and Dumas, with the hope of receiving the endorsement of reputable science. Arago offered him a better deal. In exchange for learning the secrets of the process, Arago promised to negotiate for Daguerre the purchase of the photographic process by the state.

For nearly a year, Arago mediated discussions between Daguerre and the French government, all the while leaking just enough information to keep the public's breath bated.[17] He organized a commission through the Academy of Sciences to report on the quality of his images. Arago, Biot, and Alexander von Humboldt examined previous examples of Daguerre's work, which consisted primarily of Parisian sites such as the Tuileries and Notre Dame, and witnessed the making of new images.[18]

Arago's first report to the Academy, given on January 7, 1839, was brief and enthusiastic. Daguerre, he reported, had found a way to preserve the "truth of form" of the images produced in a *chambre obscure*, by discovering a particular kind of screen on which "the optical image leaves a perfect imprint."[19] The image was reproduced "in the most minute details, with an exactitude, a fineness, that is unbelievable."[20] It had "an almost mathematical precision" that conserved exactly the photometric relations of the exterior object, even under examination with a magnifying glass.[21] Arago's thoughts turned immediately to capturing the light rays of celestial objects that he and his student astronomers spent their nights examining at the Paris Observatory. His first task for Daguerre was to record an image of the moon on his plate, and he happily reported that they did indeed come up with a visible smudge.

A preliminary interlude: Phosphorescence

Daguerre's sensitive plates promised to be an ideal detector for the elusive chemical rays. The composition of the plates, however, would remain unknown until a financial deal was struck. Physicists unwilling to wait turned thus to another range of phenomena in which light engendered chemical modifications: phosphorescence.

---

[16] Maurice Daumas, *Arago: La jeunesse de la science* (Paris: Belin, 1987).

[17] R. Derek Wood, "A State Pension for L. J. M. Daguerre for the Secret of his Daguerreotype Technique," *Annals of Science*, 54 (1997): 489–506.

[18] François Arago, "Fixation des images que se forment au foyer d'une chambre obscure," *CRAS*, 8 (1839): 5.

[19] Ibid., 4.

[20] Ibid.

[21] Ibid., 5.

In the spring of 1839, Arago began presenting work that Daguerre had completed on phosphorescence in 1824. Daguerre had, while working with phosphate powder, left a disc of blue glass on top of one of his sample dishes. He later noted that the portion of powder that had been covered by the disc glowed even more strongly than the portion that had received light directly. And thus blue light alone had a greater effect than full sunlight, from which Arago concluded that there must be rays within white light that act to impede the development of phosphorescence.[22]

When Arago finished, Biot offered an immediate retort. Daguerre had provided Biot with the same piece of glass, and Biot had carefully examined the light it transmitted. What Arago had failed to sufficiently point out, he claimed, was that what the eye perceived as blue light was in fact a complicated sum over the entire spectrum. He worked out what this sum was, and then calculated the equivalent mixture of pure blue and white light that would give rise to the same sensory perception.[23]

The argument may seem minor, but it replicated one of the central issues that had divided Arago and Biot in the past, and would continue as a point of contention in their work on photography. Where Arago simply spoke of blue light, Biot took pains to point out that what the eye saw masked a complexity in which highly different compositions could appear to be the same.

France's leading authority on phosphorescence, Antoine Becquerel, soon became involved. On February 11, 1839, he presented at the Academy a memoir on the ability of the light from an electrical charge to restore the property of phosphorescence to objects. Immediately after reading his work, he recounted, Biot approached him and suggested that the action he described was perhaps due to a form of invisible radiation distinct from that which affected the human retina.[24] Biot suggested they work together in Becquerel's laboratory at the Jardin des Plantes to examine the nature of this phosphorogenic radiation. Their experiments, explicitly based on Melloni's work with calorific radiation, used semitransparent screens to isolate the phosphorogenic radiation and establish its distinctness from visible radiation.

Biot's private invisible world

Working with Becquerel, Biot also managed to procure some photosensitive material from Daguerre. These were not the silver plates being passed around the Academy, but an alternative paper process that Daguerre claimed he had developed in 1826 and abandoned for the more promising silver plate

---

[22] Arago, "Phosphorescence du sulfite de bouyte calciné; communication de M. Arago sur quelques expériences de M. Daguerre," *CRAS*, 8 (1839): 243.

[23] Ibid., 245.

[24] Jean-Baptiste Biot and Alexandre Becquerel, "Sur la nature de la radiation émanée de l'étincelle électrique, qui excite la phosphorescence à distance," *CRAS*, 8 (1839): 223.

process.[25] As the imperfection of the paper image gave it little commercial value, Daguerre felt no need to keep it secret. It amply suited Biot's purposes, however, and he immediately replaced his phosphorescent material with sensitive paper.

Biot presented his work to the Academy of Sciences on the session of February 18, 1839.[26] The first thing he did was provide the members with a note describing the preparation that he had used, and passed around several examples that Daguerre had made up. Biot pointed out in this note that as a means of representing nature, paper had certain problems, but that as a means of detecting radiation, it was thoroughly reliable. One issue with the paper process was that it produced a reverse image of what was seen. This was, admitted Biot, a "capital inconvenience for the reproduction of nature in general."[27] But Biot's goal was not the reproduction of nature in general, but the detection of specific forms of radiation. "Little importance, indeed," he stated, "for simple physics experiments, whether the lights and darks of objects are or are not inverted, as long as the radiation's effect is manifested almost instantly."[28]

Biot's experimental arrangement was patterned directly on Melloni's work with calorific radiation. He placed semitransparent screens (colored glasses that allowed some portion of the light to pass through) between the photosensitive paper and a light source. The screens, he determined, could separate out different forms of radiation that usually traveled together. For example, light with no visible component at all could still produce a photogenic effect. His conclusion stretched beyond the simple taxonomy of light in place since Melloni:

> One is led to consider generally the radiations emanating from a body as composed of an infinity of rays, having various qualities and speeds, susceptible of being emitted, absorbed, reflected, refracted; and which, according to their proper qualities, among which one must include their nature and their actual speed, can produce vision, heat, determine certain chemical phenomena, and probably exert many other yet unknown actions, when they are received by organs sensitive to their impressions.[29]

Every body thus emitted, for Biot, a complex mixture of different kinds of radiation, which were distinct, although usually traveled together. These could be received, either separately or simultaneously, by all sorts of substances, and there produce an effect in accordance with the excitability of the recipient substance.[30]

---

[25] In fact, although Niepce had used paper as early as 1816, Daguerre most likely did not work with it before 1829. Joel Snyder, communication to author.

[26] Biot, "Sur des nouveaux procédés pour étudier la radiation solaire, tant direct que diffuse," *CRAS*, 8 (1839): 259.

[27] Biot, "Note de M. Biot sur un papier sensible préparé par M. Daguerre," *CRAS*, 8 (1839): 248.

[28] Ibid., 246

[29] Biot, "Continuation des expériences sur la nature des radiations qui excitent la phosphorescence, et qui déterminent certaines actions chimiques," *CRAS*, 8 (1839): 321.

[30] Ibid., 324.

Fig. 5.1 The device Biot referred to as a *"chambre noire."* (From Biot, *Traité.*)

The issue of resemblance was nowhere present. And it could not have been. The "photographs" Biot took were pieces of uniformly white paper, taken without a lens. He placed his device within a box with an aperture that could be opened or closed, and any handling of the paper outside of the box must be done within a specially darkened room, or what Biot called a *chambre obscure*. It is worth spending a minute on what he meant by the phrase *"chambre obscure."* For him, this was the typical laboratory space of optical experiment.[31] It was a room several meters in length, completely sealed from light except for a hole along a wall which in this case he left sealed. The key is that this hole had no lens or other means of focusing the light into a recognizable image. It produced, rather, a homogeneous beam of light whose properties could be investigated by passing it through prisms, polarizers, etc. The device with a lens used as a drawing aid was what he called a *chambre noire* (see Figure 5.1).[32]

In a general article for the *Journal des savants*, Biot reiterated the potential of the photograph to reveal the workings of an invisible world. Daguerre's invention was, for Biot, "the beginning of a new chemistry," which allowed the

---

[31] Nearly all of his work on rotary polarization was done within a *chambre obscure*. He had one in his country house at Nointel.

[32] Biot, *Traité de physique expérimentale et mathématique* (Paris: Deterville, 1816).

physicist to explore "a new class of actions, not yet sufficiently studied, which undoubtedly influence a multitude of functions of organized beings."[33] This class of actions was the effect of invisible radiations. It was realized that light had a variety of effects on substances other than the human retina. What Biot pointed out in his account was that what was commonly referred to as "light" was in fact a collection of diverse parts. Not only could white light be broken into constituent parts, each of which affects the retina differently, but the recent work of Melloni also established the existence of "specially calorific rays" that produced the sensation of heat. The rays that left an impression on silver chloride plates were a similar sort of "invisible ray" that had no effect upon our retina. Melloni had shown of calorific rays that "their existence is physically distinct, and independent, of the rays that excite in us the sensation of light," and that each could be isolated as to have the effects of heat without the effects of light and vice versa. It remained an issue of great interest, claimed Biot, whether a similar independence could be established for chemical rays (and this very issue formed the crux of his own current research interests).

Biot emphasized the mystery of a world inundated with a potentially infinite number of diverse forms of radiation, most of which remained inherently unknowable to man. The photograph was a valuable tool because it opened up a corner of this world for investigation; it allowed the observer access to a range of phenomena that had previously been denied to him. Like his polarimetric traces revealing the "vital forces" of living matter, Biot's photographic papers were less representation than a form of communion with a world beyond.

Arago's collective visibility

If Biot's solitary withdrawal into his darkened chamber replicated his optical practice of the previous twenty years, then Arago's own creation of a team of junior daguerreotypists was no less situated in the practice he had instilled at the Observatory throughout the 1830s. Arago had, following his appointment as Director, crafted the Observatory into a machinery of visibility from an aggregate of observers, instruments, and instrument makers. The product of their surveillance, the astronomical observation, was a pure visual image whose epistemic status was rooted in the transparency of the operations that produced it. The daguerreotype, in 1839, became another piece of this carefully managed organization of observation.[34]

[33] Biot, "Sur les effets chimiques des radiations, et sur l'emploi qu'en a fait M. Daguerre, pour obtenir des images persistantes dans la chambre noire," *Journal des savants* (April 1839): 174.

[34] Astronomical photography has received particular attention by historians of science, see, for example, G. Holmberg, "Mechanizing the astronomer's vision: On the role of photography in Swedish astronomy, c.1880–1914," *Annals of Science*, 53 (1996): 609–616; John Lankford, "Photography and the Nineteenth Century transits of Venus," *Technology & Culture*, 28 (July 1987): 648–657; Alex Pang, "'Stars should henceforth register themselves': Astrophotography at the early Lick Observatory," *British Journal for the History of Science*, 30 (1997): 177–202; Alex Pang, "The Social Event of the Season—Solar Eclipse Expeditions and Victorian Culture,"

One of the first scientific applications Arago suggested for the daguerreo-type was photometry.[35] The measurement of light intensities was a long-standing problem in astronomy, and one particularly close to Arago's heart. He had, throughout the 1830s, designed several photometers and written at least seven memoirs on the subject.[36] His aim was to come up with a reliable device that his *élèves astronomes* could use to give readings interchangeable with one another. Photography seemed to offer an ideal solution. Arago thus proposed a set of experiments, modeled after those conducted with his own visual photometer, that replaced the human eye with a sensitive plate. He enlisted two young daguerreotype enthusiasts, Hippolyte Fizeau and Léon Foucault, to conduct the experiments for him.

Fizeau and Foucault were both twenty years old in 1839. They lived around the corner from one another, Fizeau at 17 rue de Cherche-Midi and Foucault at 30 rue d'Assas. They became, along with several other young men of the neighborhood, impassioned by the practice of photography.[37] "L'ecole de Saint-Sulpice," as they have been called, specialized in images of the proliferating rooftops of urban Paris. Resembling the views from their garret apartments, these photographs were a far cry from the pastoralism of Talbot's Lacock Abbey. Fizeau and Foucault, in turn, were very different from the well-trained savant that Biot hoped would serve as the gatekeeper to the mysterious world of the photographic image. Both, indeed, were school dropouts who seemed to have a knack of working with their hands. Their strategies for investing the photograph with reliability were equally different.

Although neither of them officially joined the staff of the Observatory, they did much of their work there. Their most celebrated collaboration, the 1850 comparison of the speed of light in water and air, was an experiment that Arago had drawn up several years before, but had never conducted because his eyes were too weak.[38] Their most noted individual accomplishments, Fizeau's measurement of the speed of light and Foucault's famous pendulum, also took place within the physical space of the Observatory.[39]

Isis, 84 (1993): 252–277; Holly Rothermel, "Images of the Sun: De La Rue, Airy and Celestial Photography," *British Journal for the History of Science*, 26 (1993); Simon Schaffer, "Where Experiments End," in *Scientific Practice*, Jed Buchwald, ed. (Chicago: University of Chicago Press, 1995).

[35] For example, in his speech before the Chamber of Deputies: Arago, *Rapport sur le daguerreo-type* (Paris: Bachelier, 1839).

[36] Reproduced in Arago, *Oeuvres complètes de François Arago*, J.A. Barrel, ed., 12 vols. (Paris: Gide et J. Baudry, 1854), vol. 10, 152–298.

[37] Fonds Fizeau, Archives de l'Académie des Sciences.

[38] Foucault, *Sur les vitesses relatives de la lumière dans l'air et dans l'eau* (Paris: Bachelier, 1853).

[39] Foucault in fact first conceived of the pendulum while working on the clock mechanism for moving telescopes. Foucault, *Recueil des travaux scientifiques; publié par Madame Veuve Foucault sa mère, mis en ordre par C. M. Gabriel* (Paris: Gauthiers-Villars, 1878). For more on their independent careers, see Jan Frercks, *Die Forschungspraxis Hippolyte Fizeaus: Eine Charakterisierung ausgehend von der Replikation seines Ätherwindexperiments von 1852* (Berlin: Wissenschaft und Technik Verlag, 2001) and William Tobin, *The Life and Science of*

The photometric work that Fizeau and Foucault undertook in the 1840s was patterned on a set of experiments that Arago had performed in the 1830s, which consisted of comparing the light of the sun to that produced by burning lime in a hydrogen flame. Instead of a visual photometer, however, they used photosensitive plates to record the radiation.[40]

Even when the photosensitive plate did not agree with the photometer, it was still given a privileged relationship with respect to the eye. In a treatise on photometry written near the end of his life, Arago took up the question of whether or not the edges of the sun possessed the same brightness.[41] At stake was the physical constitution of the sun: a brighter center was consistent with a solid or a liquid, while a uniform intensity implied a gas. Arago began by providing a history of the problem. Galileo and Huygens had concluded, from theoretical suppositions, that there was no difference in intensity. Bouguer, however, "the first experimentalist to pronounce on the question," found a difference between center and edges on the order of 45:38. Physicists remained divided, with Lambert claiming no difference, and John Herschel, Airy, and Laplace seeing one. Arago applied a technique of his own to the problem that relied not on "mere estimations," but on "measurements."[42] He used a polari-scope to divide the sun's image into two discs of complementary colors. He then superimposed the edge of one image with the center of the other. The fact that there was no perceptible color implied that the two images had intensities that differed by no more than 1/40th, a figure Arago had established in previ-ous work as the limit of the sensibility of the eye. Accordingly, Arago gave the ratios as 40:41, and declared Bouguer's 45:38 "inadmissible."

"The study of light's photogenic effects," he claimed, "would also help resolve this question." Immediately following Daguerre's announcement, Arago reported, he had formed an image of the sun on a silver plate, and found that the photogenic effect was significantly greater in the center than on the edges. Foucault and Fizeau continued this project of photographing the sun, producing several images of the object in the years 1844 and 1845. They, too, Arago reported, were struck by the same observation.[43] These images, some of which found their way into Arago's textbook on popular astronomy, were admirably sharp. Sunspots with a diameter 1/200th that of the sun were easily visible. "Thus it does not appear possible to attribute the decrease in intensity

---

Léon Foucault: The Man Who Proved the Earth Rotates (Cambridge: Cambridge University Press, 2004).

[40] Léon Foucault and Hippolyte Fizeau, "Recherches sur l'intensité de la lumière émise par le charbon dans l'expérience de Davy," *CRAS*, 18 (1844): 746–754; Fonds Fizeau, Archives de l'Académie des Sciences.

[41] Arago, "Quatrième mémoire sur la photométrie: constitution physique du soleil, lu à l'Académie des Sciences, le 29 avril 1850," *Oeuvres complètes de François Arago*, J.A. Barrel, ed., 12 vols. (Paris: Gide et J. Baudry, 1854), vol. 10, 231.

[42] Ibid., 234.

[43] They first reported the observation in conjunction with their 1844 photometric work. H. Fizeau and L. Foucault, "Addition à une précédent Note concernant l'application des procédés daguerriens à la photographie," *CRAS*, 18 (1844): 860.

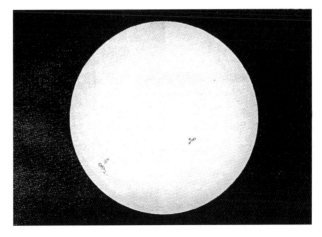

Fig. 5.2 Foucault and Fizeau's 1844 daguerreotype of the sun. (From C.M. Gabriel, *Recueil des travaux scientifiques de Léon Foucault* (Paris: Gauthier-Villars, 1878), pl. I.)

to a defect in the sharpness of the images," Arago concluded.[44] Rather, it must be attributed to a difference in the properties of the rays coming from different parts of the sun. The chemical actions of the rays from the center were noticeably more powerful than those coming from the edges (see Figure 5.2).

Arago gave a short account of his work in one of his lectures on popular astronomy soon after:

> This is not the place to enter into details on the experiments which I have executed on this subject, from which I concluded that there is a difference in intensity between the edge and center equal to 1/40. [...] Two very distinguished physicists, MM. Fizeau and Foucault, having produced, on my request, a very rapid impression of the disc of the Sun on daguerrian plates, have verified by photography the results to which I have been led by photometry.[45]

And thus Arago was able to claim that his visual measurements, which showed no difference in intensity, were verified by the photographic images of the sun, which showed a clear difference!

In question was the photograph's status as inscription. What was it that served to trace the image back to the reality it was supposed to represent? When was this trace broken? These were questions decided at the level of technique. For Arago, daguerreotypes alone possessed the geometrical correspondence to the visible world that ensured their reliability. They alone withstood the *"l'épreuve de la loupe,"* in which the lines retained their distinctness and

[44] Ibid., 248.

[45] Arago, *Astronomie Populaire*, J. Barruel, ed., 4 vols. (Paris: Gide et J. Baudry, 1854–7), vol. 2, 169–170.

accuracy even under a magnifying glass. In the first photograph made of an astronomical body, Fizeau and Foucault's daguerreotype of the sun, Arago claimed that one could use it with confidence because of the "*netteté*" of the image and the appearance of sunspots that could be seen with the eye. By focusing on the issue of resemblance, moreover, Arago could use pictures that he himself had not been involved in making. Biot, on the other hand, was involved in every step of the process. It was his knowledge of how this process occurred that assured him the surface and showed what he thought it did. The "*netteté*" of the image was simply not a concern. Biot was, after all, not even using a lens to focus the image. The porous surface of paper, therefore, was entirely acceptable, and even preferable, due to its more basic chemistry.

For Arago, the invisibility to the human eye was neither reason to think that the phenomenon did not exist, nor that the eye and the photograph saw different things. Rather, it was a demonstration of the utility of the photograph in extending the limited range of human vision. Not only did the photograph and the eye see the same thing, but it assured for Arago that the eye was seeing something, even if it was below the threshold of its sensitivity.[46]

At the end of his memoir on photometry, Arago added that to complete the study, one would need to examine whether the spectrum from the center of the sun contained dark lines at the same location as the spectrum from the edges. The idea was to provide a test for the identity of the two types of radiation. It was one of the first techniques that Arago had drawn on to establish that visible and chemical light were the same thing. He first brought up the idea in his speech of August 19, before the Academy. Observation had shown, he said, that the solar spectrum was not continuous, but contained several entirely black lines. "Are there similar discontinuities in the dark rays that appear to produce photogenic effects?," he asked.[47] He outlined how one would answer it. The discontinuities of the visible spectrum were perceptible to the naked eye. To render photogenic effects perceptible, "one would make a sort of artificial eye by placing a lens between the prism and the screen where the spectrum falls."[48] The lens would thus simulate the lens of the eye, and the photosensitive screen would be the retina. One could then look, perhaps even using a magnifying glass, at the black lines that form on the screen.

Arago did not carry out the experiment himself. The Observatory's *chambre obscure* was being moved, he explained, and its replacement had not yet been completed. He did get his answer, though, when the young physicist Edmond Becquerel completed a memoir on the subject in 1842. He reported that, when casting a solar spectrum on an iodated plate, one did indeed observe

---

[46] John Tresch quite correctly points to the many ways Arago's daguerreotypes fail as truly perfect imitations of the world, yet this does not negate Arago's use of resemblance as a strategy to ground the daguerreotype's status as representation. John Tresch, "The daguerreotype's first frame: François Arago's moral economy of instruments," *Studies in History and Philosophy of Science*, 38 (2007): 445–476.

[47] Arago, "Le daguerreotype," *CRAS*, 9 (1839): 264.

[48] Ibid.

lines of unmodified chemical matter at exactly the same locations as the black lines of the luminous spectrum.[49] Arago presented the memoir to the Academy on June 13, 1842.

Arago used the results to get in a dig at Biot and his world of mysterious dark rays. To most people, Arago admitted, the experiment looks rather superfluous. Would not it be obvious that there would be no photogenic action where there was no light? Yet Arago recalled that not everyone accepted that the modification of impressionable substances was caused by the action of solar light itself. "Some people" preferred to think of the modifications as caused by a sort of dark radiation mixed in with light, traveling with it, and undergoing the same refraction. If this were the case, what the experiment actually showed was that not only was the spectrum of these invisible rays not continuous, but also it had its discontinuities in precisely the same place as the visible spectrum. "That would be," he said, "one of the most curious, one of the strangest results of physics."[50]

The case of color

Arago raised the question, in his famous speech before the Academy of Sciences, "Will this process be susceptible to perfection? Will one come to produce colors?"[51] The standard answer, Arago admitted, was no. But he was more hopeful. He cited a suite of previous work that had given promising results. Daguerre had reproduced some colors while working with phosphorescence. Niépce had often been struck by what he believed to be colors. John Herschel had obtained different colors while viewing the solar spectrum. Seebeck, using silver chloride, was able to produce the color violet by the incidence of violet light, as well as blue by blue. Interestingly, Talbot, who had done perhaps the most work in this area, went unmentioned by Arago. No one had yet obtained a pure red, Arago admitted, "But all the same, these attempts are far from being discouraging."[52] He placed his faith in the future work of physicists, which would undoubtedly place these first efforts on firmer ground. "In the presence of these facts, it would certainly seem foolhardy to assert that the natural colors of objects will never be reproduced in photogenic images."[53]

Biot was of a different opinion. He discussed, as Arago had, Daguerre's early attempts to produce color with phosphorescence, but with a decidedly less optimistic tone. He stressed that there was no necessary reason any substance should emit the same kind of radiation that excited it. As Biot pointed out, "experiments done since long ago by physicists have shown that there is no

[49] Edmond Becquerel, *CRAS*, 13 (June 1842).

[50] Ibid.

[51] Arago, *Le Daguerreotype: Rapport fait à l'Académie des Sciences de Paris le 19 Août 1839*, (Paris: Bachelier, 1839), 24.

[52] Ibid., 24.

[53] Ibid.

such correspondence between the color that the affecting rays produce in our eye, and the color of the light emitted by the affected powder."[54] And he took Daguerre's failure to find such a substance as conclusive proof that it did not exist. He described the work as "an infinity of experiments on phosphoric powders in general," implying that no further experimentation could be done (far different from Arago's announcement that Daguerre's experiments with phosphorescence "have given a few results"). Finding a substance that represented objects by their natural color would thus be nothing more than coincidence. Far from viewing Daguerre's work as a promising beginning, he saw the failure to reproduce more than one color at a time as a demonstration and that it was "impossible to obtain in general" an agreement between emitted and incident light.[55]

Biot also made it clear that photogenic color was equally impossible. Again, the problem was the lack of a necessary connection between object and image. When a painter wanted you to see a colored image, he arranged coloring substances on a surface such that the light sent back to your eye was dominated by the particular rays that formed the tint he wanted you to see. "But the chemically active radiation that the same parts of the picture receive and give off is distinct from the light that affects your retina." This was Biot's key point, and he invited the reader to refer to his experiments of 1839, which established this distinction as a fact.[56] To get natural colors, he continued, two things needed to happen. First, the radiation reflected by the surface had to be chemically active. Second, the energy of its action had to be proportional to the intensity of illumination affected in the eye by the luminous radiation from the same point on the surface. But this second agreement was unlikely. There was a large number of materials with completely different chemical compositions that nonetheless affected the eye in a similar manner, as painters knew who freely substituted the one for the other. Two substances of the same color could be as different in the invisible radiation they emitted as two substances of different colors. The inverse could also be true: substances which were similar in their property of reflecting chemical radiation could be quite dissimilar to the eye. "These are difficulties generally inherent in the making of chemical pictures," Biot concluded, "and they show, I think, quite obviously, the illusion of the experimenters who have hoped that one will be able to match up, not just the intensity, but the colors of the chemical impressions produced by radiation, with the colors of the objects from which the radiation is given off."[57]

Biot performed some work of his own on gaiac resin to demonstrate his point. This substance was well known as a curiosity by physicists. When a piece of paper treated with it was exposed to light on the blue end of the spectrum, it turned blue. When it was exposed to light from the yellow end of the spectrum, it turned yellow. This was indeed striking. "But," warned Biot, "the effects

[54] Biot, "Sur les effets chimiques des radiations," 202.
[55] Ibid.
[56] Biot, "Notes sur des dessins photogeniques de M. Talbot," *CRAS*, 10 (1840): 485.
[57] Ibid., 486.

thus observed are complex and mixed with an optical illusion."[58] The resin was in fact chemically composed of two different substances. One of them was yellow, and was unaffected by exposure to radiation. The other had at first no color, but became blue on exposure to light. And thus, two completely different chemical processes were responsible for the different colors. Beneath the illusory surface resemblance lay a complexity only revealed through the subtlest of manipulations.

The question of the possibility of color reproduction would be largely resolved by Edmond Becquerel in 1849. The younger Becquerel, the son of Antoine, moved back and forth between paper and silver plate techniques, between claims of chemical and optical reliability, and in and out of realist claims for the photograph. The trajectory of his work shows a concern with rooting his representational claims in laboratory technique. I want to follow two episodes in it. In the first, the distinction of the *rayons continuateurs* from the *rayons excitateurs*, he emphasized the centrality of chemical complexity. In the second, color photography, he puts these considerations aside to focus on the accurate reproduction of what he sees.

In 1839, Edmond Becquerel was nineteen years old. He worked with his father at the Jardin des Plantes, and served as an assistant at the Faculté des Sciences, where Biot was a professor of physics. He began his work immediately after the impact of Daguerre's announcement began to be felt, and his first duties included the preparation of sensitive paper for his father and Biot. In his 1839 work on the separability of visible and chemical light, Biot acknowledged the younger Becquerel as "an assistant as intelligent as he is zealous."[59]

As Edmond Becquerel assisted in the work of his senior physicists, he also conducted research toward his thesis in physics at the Faculté des Sciences. His project, entitled *Des effets chimiques et électrique produits sous l'influence de la lumière solaire*, was another chapter in the effort to establish the physical independence of chemical radiation.[60] His first sentence reiterated the standard line: "Solar radiation is not composed solely of luminous rays perceptible by the organ of vision, but also of rays of different orders endowed with particular properties."[61] His aim was to arrive at a way of perceiving these "other rays" contained in solar light. He constructed a device with which one could measure the intensity of the chemical reactions caused by sunlight on metal plates. It consisted of plates of two different metals placed within a conducting fluid. He then permitted sunlight to fall on the plates. If photoreactive, the plates would be chemically altered, and the

---

[58] Biot, "Sur le pouvoir de la radiation atmosphérique comme agent chimique," *CRAS*, 8 (1839): 599.

[59] Biot, "Sur des nouveaux procédés," *CRAS*, 8 (1839): 260.

[60] E. Becquerel, *Des effets chimique et électrique produits sous l'influence de la lumière solaire: Thèse de physique presentée et soutenue à la Faculté des Sciences le Aôut 1840* (Paris: Firmin Didot, 1840).

[61] Ibid., 5.

electric current between them would change. This current could be measured by sticking two electrodes at opposite ends of the fluid bath, and, Edmond Becquerel claimed, represented the intensity of the chemical radiation. Using semitransparent screens to isolate different kinds of radiation, and his newly constructed device to detect their chemical effects, he presented his work as a means of seeing that part of sunlight invisible to the eye.

The work fit into the research efforts Biot had called for in his *Journal des Savants* article. Edmond Becquerel's thesis cited him more than any other physicist and bore his signature as the Doyen of the Faculté des Sciences. But there was one crucial point that Biot was not at all happy with. Edmond Becquerel's device was indeed ingenious, he said, as an indicator of the presence of chemical radiation.[62] But that did not necessarily mean it could be used to measure the amount of chemical radiation present. Effects were not always strictly proportional to their physical causes. If this cause was complex, as was certainly in the case with radiations, it was in fact unlikely that this proportionality could be invoked. Diffuse radiation contained within it elements that were capable of acting on certain substances in opposite ways, such that light could sometimes have an even greater chemical effect after it had lost some of its rays in a semitransparent screen. Which meant there was no simple relation between the effect and the number of rays.

As Edmond Becquerel was not present to defend himself, his father got up and did it for him. There was only one chemical reaction occurring, he pointed out, and it was thus "useless to worry about the different chemical radiations that can exist in light."[63] Any rays producing opposite effects would give rise to a current in the opposite direction, and thus would be measurable. Biot responded that this did not seem at all to respond to his objection that the effect was not necessarily proportional to the total number of rays. Even his closest student did not seem cautious enough in guarding against the tricks radiation could play in making itself visible.

Edmond Becquerel also used sensitive paper to investigate the properties of chemical light. He concluded from this work that chemical radiation was composed of at least two different orders. The first, which he named *rayons excitateurs*, had the property of instigating a chemical reaction. The second, which he named *rayons continuateurs*, could not themselves start a reaction, but could continue the action of other rays. Edmond Becquerel projected a solar spectrum on to a paper treated with silver bromide. Normally, the paper would only become darkened where the blue and violet rays were hitting it. If, however, before exposing the paper to the spectrum, one let it be slightly affected by diffuse light, then one would find that the paper would become colored not only in the blues and violets, but also in the greens, yellows, and all the way up through red. This coloration in the less refrangible part of the spectrum was due, he concluded, to rays that acted differently from those that

[62] Biot, "Sur des nouveaux procédés," 170.
[63] Ibid., 173.

had already been studied. He investigated the effect further by preparing a sheet of paper covered with parallel bands, of which every other one had been exposed to diffuse light. When a solar spectrum was then cast on the paper, it blackened the previously unexposed bands only in the region above the blue, whereas it blackened all of the previously exposed bands. He then went on to establish the nature of the radiation using the technique he was already so well acquainted with from the work with his father and Biot: semitransparent screens.

This same phenomenon came under study by the silver plate partisans some time later. On September 28, 1846, Arago reported to the Academy an observation made by the optician of the Observatory, Noël Lerebours. He had been working to construct objective lenses to use while taking daguerreotypes, and had noticed that images formed with white light were less striking than images formed with blue light alone. From this he concluded that the less refrangible rays (greens, yellows, oranges, and reds) seemed to inhibit the action of rays on the other end of the spectrum.

Within a few days, Foucault wrote to Arago informing him of some results he had found in conjunction with Fizeau.[64] The two of them had prepared a silver plate in the typical manner, then exposed it to the light from a lamp for an amount of time sufficient to begin to see an alteration. Next, before exposing it to mercury fumes, they let a solar spectrum fall on it for a set amount of time. They then placed the plate in mercury, and examined the result. They found that the rays starting with orange and going up to violet left a white impression that stood out against the gray background of the plate. The rays of lesser refrangibility, however (i.e. the reds and even slightly beyond), left a very dark impression, where the silver coat was not at all affected. The first impression, from the lamp, had been "destroyed or neutralized" by the red rays. Fizeau and Foucault concluded that the solar spectrum could be divided, with regard to how it behaved, into two parts: one with a "positive action" that served to increase the intensity of the impression of a previously affected plate, and the other with a "negative action" that served to diminish it. "Negative action" was the term reserved for the extreme red end of the spectrum, whereas "positive action" described the rest of the rays, "on the effectiveness of which rests the entire photographic art."[65]

Edmond Becquerel responded that this was not at all the case. The two men, he claimed, did not fully appreciate the chemical complexity of the situation. He warned: "One must therefore be on guard against the appearances presented by the deposits formed on the surface of daguerrian plates; and, if one has but these facts alone to affirm the existence of rays acting in various ways, one runs the risk of being mistaken."[66] He cited the example

---

[64] Foucault, "Observations de MM. Foucault et Fizeau concernant l'action des rayons rouges sur les plaques daguerriennes (Lettre de M. Foucault à M. Arago)," *CRAS*, 23 (1846): 679–682.
[65] Ibid., 680.
[66] E. Becquerel, "Observations sur les expériences de MM. Foucault et Fizeau relatives à l'action des rayons rouges sur les plaques daguerriennes; par M. Edmond Becquerel," *CRAS*, 23 (1846): 800–804.

of an experiment with a daguerrian plate where, if one shone a spectrum on the plate for a short period of time, one saw black Fraunhofer lines on a white background, but if one let the action continue for an hour or more, one saw white Fraunhofer lines on a black background. These "inverse effects," which were precisely the same effects that Foucault and Fizeau were witnessing on their plates, were "secondary effects produced by several chemical reactions operating simultaneously."[67] He cited an observation made by John Herschel, another researcher who used almost exclusively paper. In this case, however, Herschel had treated the paper first with a solution of lead acetate, and then with a bromide of potassium and silver nitrate. He had obtained a spectrum that whitened in the blues and darkened in the reds. It was not at all a question of two distinct effects, positive and negative, produced by the radiation from a single material. Rather, it was a question of two different chemical reactions. First, the light affected the silver iodide, which brought about coloration. Then, the light caused the decomposition of the potassium iodide, and the iodination of the silver that was produced in the first reaction. The *rayons continuateurs* contained in the red part of the spectrum prolonged the first reaction, once started, while the blue end was dominated by the second reaction. This observation, Edmond Becquerel pointed out, showed that a mixture of sensitive material could give rise to a number of different reactions, and when one looked at the photographic plate one saw only the final result of a complex process.

It was for this reason that he stated that "one must only use daguerrian plates as little as possible when doing research on the nature of active rays." These plates consisted of iodinated silver that had been exposed to vapors of bromine or chlorine. The silver salts under study (silver iodide, etc.) existed in the presence of metallic silver. When the action of light decomposed the salts into subsalts and free iodine, chlorine, and bromine, the possibility of reacting with metallic silver meant that the process, "already very complex, become even more so."[68] The mistake of Foucault and Fizeau was to try to use the daguerreotype as "a particular sensitive surface" for studying light when in fact it was a mixture of materials that behaved differently in different parts of the spectrum. Daguerrian plates, said Becquerel, "as important as they are for the photographic representation of camera obscura images" were worthless for studying the chemical action of rays.

Yet as he wrote this, Edmond Becquerel was also engaged in a series of researches involving daguerrian plates. In this case, however, he departed from the strict goal of investigating the properties of unseeable radiation, and occupied himself with the reproduction of visual effects. His goal was the very "illusion" that Biot had dismissed as fanciful, that is, the reproduction of images in their natural color.

[67] Ibid., 801.
[68] Ibid., 803.

Edmond Becquerel was interested in color photography from the moment of his work with photosensitive materials in 1838.[69] He was well aware of the work done by Seebeck and Herschel, and although he followed Biot in assuming that the effects of color were due to different chemical reactions, he recalled nonetheless that "it was a rather curious coincidence to see the two extremities of the photogenic impression of the spectrum on the silver chloride turn, the one violet in prismatic violet, the other red in prismatic red."[70] Throughout the 1840s, he worked on ways to get the best effects out of silver chloride. His first efforts were with paper, but he found he got better results with silver plates. He came upon the following process in 1848. He first exposed a silver lamina to chlorine gas, whereon it became a grayish-white. Then he projected a spectrum on it, which led to a slight darkening in the violet region. Next he immersed the lamina into chlorinated water for several instants, where it became covered with a grayish-white coat. This time, projecting a spectrum on to it caused faint but correctly nuanced colors to appear.

"I saw then," he wrote, "that it was not a simple coincidence of color that had given the previously exposed silver chloride the color red on one end of the spectrum and violet on the other."[71] Rather, it must be the case that a subchloride was being created, and mixed with the original unaltered chloride, and that it was this new substance that was giving rise to all the various tints. Here, then, was the key to color photography: a single substance that reproduced an object's real colors. This process never became entirely viable, as Becquerel could not find a way to make the colors stable. But it did restore confidence that color was not an unreasonable expectation.

At the very moment that Edmond Becquerel came to believe that the color was not due to coincidence, that is, that it corresponded to something out in the world rather than in the chemistry of the process, he switched from paper to metal plate. He would in the following months return to paper, in an attempt to reproduce colored images on this surface. But ultimately, he would abandon this work as unpromising, claiming the results were consistently "*moins beaux*" than those produced on plates.[72] And thus, in the end, it was the beauty of the effects that persuaded Becquerel, and drew him from the materials of physics experiments to those of art.

For Edmond Becquerel, the switch in epistemic status coincided with his switch in practical technique. When he was interested in the properties of the mysterious and invisible *rayons continuateurs*, he used paper. He was then able to use the straightforward chemical nature of the paper prints to argue that his images were more reliable than those of Fizeau and Foucault. When he became interested in recording the manifestly visible colors of the spectrum, he found he was more satisfied with silver plates. He then cited the likeness of the colors to the spectrum as evidence of their reliability. In this way,

[69] E. Becquerel, *La lumière, ses causes et ses effets* (Paris: Firmin Didot frères, 1868), 210.
[70] Ibid.
[71] Ibid., 211.
[72] Ibid., 217.

meaning and technique reinforced one another. It was the object under study, its visibility or invisibility, that determined the practical technique. But then the technique, in turn, was used to construct the photograph's status.

For Biot and Arago, both meaning and technique were tied up in the concerns of their distinct activities. Arago and the astronomers were interested in the light given off by stars. They transformed Arago's visual photometer into a photochemical one while maintaining the bulk of their experimental procedure. Biot, on the other hand, was interested in the interactions of radiation and matter. He, too, inserted the photograph into a previously existing laboratory tradition, this one involving separating forms of radiation and examining their effects. The revolutionary introduction of photography into scientific practice was thus accompanied by a strong continuity in materials and manipulations. And it was on the level of materials and manipulations that the experimenters involved decided what the photograph was. Yet, as we have seen in previous chapters, these sets of practices were hardly politically neutral. Rather, Arago and Biot's efforts to establish the proper lines of visibility were simultaneously constrained by and constitutive of the web of power relations in which they worked.

A correspondence between gentlemen

The question of the laboratory floor: "who could produce and interpret photographic images?" was also a question for French society writ large: "who was qualified to practice the activity of science?" And science here was no cordoned-off, isolated enterprise. It was the ability to see correctly and to make judgments. It was also the ability to produce representations and to participate in the sphere of rationally communicating individuals. The question, in short, was "who made up the French public?" The answers of Arago and Biot drew the lines of civic participation in very different places.

Few people in France shared Biot's vision of a mysterious world of invisible radiation. And so he looked, as he had so often done before, to England. William Henry Fox Talbot had been working on his own process for fixing the images of the camera obscura since 1834, and had been prompted by the January 7 announcement to write a letter to Arago and Biot, claiming formal priority.[73] Both men quickly rejected this claim. But what they did not agree on was the potential utility of the process. Arago showed little interest in Talbot's work. It was, he pointed out, the same method that Niépce and Daguerre had tried long ago and given up in favor of the "much more perfect" method now employed.[74] Biot, on the other hand, was more encouraging. "Perfection is the concern of art," he wrote to Talbot after Arago's pronouncement, "but for

---

[73] Larry Schaaf, *Out of the Shadows: Herschel, Talbot, and the Invention of Photography,* (New Haven: Yale University Press, 1992).

[74] Arago, "Fixation des temps de la chambre obscure," *CRAS,* 8 (1839): 208.

physics, the sensitive papers, both his [Daguerre's] and yours (which I shall be using myself), are of the greatest value."[75]

Biot became Talbot's primary advocate in France. He would read Talbot's letters to the Academy of Sciences and pass around examples of his work. Before the first display of Talbot's photographs, Biot described to his audience exactly what it was they were getting, and how they should look at the images before them. "One should not expect that photogenic drawings, done on paper, could ever equal the sharpness (*netteté*) and fineness (*finesse*) of those obtained on smooth, polished metallic plates."[76] The spongy texture of paper, the roughness of its surface, and the effects of capillary action presented serious obstacles to "the absolute rigor of the linear trace." But this was not a serious fault for Biot, as the trace between visible object and image had been broken from the start. As long as one did not claim to be doing art, sensitive paper was "perfectly sufficient" and even better suited for the tasks envisioned for photography. If one wanted to obtain an accurate copy of a rare manuscript that could then be circulated, or record the impressions of a voyage to a distant location, paper presented obvious advantages. Several reproductions of the original drawing could be made immediately. Four or five hundred paper images could be placed in a portfolio and transported with little difficulty. The metallic plate of the daguerreotype, on the other hand, was not only itself a heavier item, but required a glass encasing to preserve the images. The elaborate precautions required of the daguerreotype led Biot to doubt the safety of these "frail products" in the course of travels that could be long, difficult, and even perilous.

Among the photographs Biot presented were several examples of exotic and rare documents rendered available for the first time. A Hebrew psalm, a Persian gazette, and a Latin charter from 1279 were among the items reproduced for the Academy of Sciences. Biot also presented them to the Académie des Inscriptions et Belles-Lettres for examination, thereby emphasizing, not their value as scientific curiosity, but as important documents, conveniently transported from their distant locations. Biot was, incidentally, admitted to this Academy the next year, in 1841, on the basis of his expertise in the science of ancient Egypt and Mesopotamia. The literary members of the Académie des Inscriptions et Belles-Lettres pronounced that Talbot's paper copies had been entirely successful in their displacement across the ocean; they were "as readable as the original texts."[77]

It was to Biot that Talbot confided the details of his process in a letter of February 20, 1839. Biot chose to open the sealed letter at the Société Philomatique, rather than the Academy of Sciences.[78] This allowed the process

---

[75] "Biot to Talbot, 24 February 1839, 1937–4851," Archives of the National Museum of Film, Photography, and Television.

[76] Biot "Notes sur des dessins photogeniques de M. Talbot," *CRAS*, 10 (1840): 484.

[77] Ibid., 485.

[78] "Correspondance, seance of 25 feb," *CRAS*, 8 (1839): 303.

to ultimately find publication in the *Bulletin de la Société Philomatique*, on whose board Biot sat, rather than the Arago-controlled *Comptes Rendus*.

Biot also kept Talbot abreast of developments in France, particularly his own work establishing the distinction between visible light and the radiation affecting photographic plates. We live, he summarized, in "a general complex radiation composed of distinct parts which are congenerous and mixed in variable proportions in the emanations from different bodies."[79] There was not one kind of light, but several. Which, he pointed out, made the word "photography" something of a misnomer. He suggested that "*actigénie*" would more fully captured than the complex nature of the "light" in question. "But," he hastened to add, "it is for those who discovered it to find a name, as they must know that I would not wish to infringe their rights of discovery."[80] Talbot took the advice to heart. In subsequent works he spoke of the "actinal rays" responsible for the images on photographic plates, and pointed to their fundamentally invisible nature.[81]

Biot and Talbot remained in close correspondence. A few months later, Talbot wrote to Biot asking him to report to the Academy that he had found a way to heighten the sensitivity of his papers by at least a hundred times. As he reported in the letter that Biot read out loud, he had been able to fix the images of the *chambre obscure* in as little as eight seconds. And that, he reminded everyone, was in October; the summer sun should allow for an even shorter exposure. Biot responded by outlining a set of experiments for Talbot to try with his improved materials. Edmond Becquerel had recently established the fact that different kinds of radiation were required to begin a reaction and to continue it once it had started. The practical implication of this instantaneous action of the radiation was that one was not required to wait until an image appeared while taking a photograph. Rather, one could expose it briefly, and then take it home and finish the process at one's leisure. Talbot wrote back that this phenomenon was quite well known to him and that it worked perfectly well with his improved paper. One could expose a piece of treated paper in a camera obscura for a few instants and then take it out. Even though there had been no appreciable change, he said, "the picture already exists in all its perfection, but in a state of complete invisibility."[82] One could wait for weeks or months, and then make the picture appear, "as if by magic" by exposing it to radiation from the red end of the spectrum and below.

Watching the image emerge from invisibility was, in Talbot's words, "Quite the most marvelous thing that one can see; and the first time I saw it I was seized by a sort of astonishment." Here was yet another distancing of the photographic record from the visible object. Now what actually traveled was what looked like an empty piece of paper. The image only appeared when

---

[79] "Biot to Talbot, s.d.," NMFP&T Archives.
[80] Ibid.
[81] Talbot, *The Pencil of Nature* (London: Longmans, 1844).
[82] "Nouveaux détails sur les papiers impressionables communiqués à M. Biot par M. Talbot," *CRAS*, 12 (1841): 226.

exposed to a form of radiation that had no relation whatsoever to the light given off by the object. Talbot saw the possibility of using the photograph as a form of invisible writing. If it fell into the wrong hands, not only it would appear to be simply blank paper, but also it would be destroyed on the first contact with sunlight. This idea, he felt, was sure to delight "diplomats … and those who love mystery."[83] And few people loved mystery more than Biot. Reproductions of arcane languages and secret spy messages were among the least obscure objects of his interest. In Biot's correspondence with Talbot as in his own work, the issue of mystery brought with it another question: did everyone have access to the secrets of nature or just the initiated few?

Biot found in Talbot a correspondent who shared not only his taste in photographic materials, but seemingly his entire vision of scientific practice. Talbot, like Biot, preferred to work in pastoral quiet of his country home. Many of Talbot's first images were of Lacock Abbey itself, an impressive estate several hours from London. These were among the first photographs that Talbot sent to Biot, who responded enthusiastically:

> I, too, Sir, am fortunate enough to have a country home, but it is a small house with a farm that I cultivate myself, or, to tell the truth, I cultivate with my eyes, far from the noise of the town and the literary life. I have horses and cattle with *English* stabling and all this in a charming situation on the road from Calais to Paris.[84]

Biot delighted in sharing his work with another country gentleman. Talbot's own practice fit well with the epistemology of pastoral withdrawal that Biot had articulated in his work on optical activity. For it was only "far from the noise of the town" that one could apply the new techniques of photography without running the risk of descending into public spectacle or showmanship. Daguerre, Biot implied, had overstepped this boundary and tainted his work with an excessive interest in its visual impact. Biot had often expressed disappointment that Daguerre had completely abandoned the "less clear prints" of paper. On one occasion, he was able to convince Daguerre to investigate the actions of some papers, but then became angry that Daguerre would not publish his work. The reason, he explained to Talbot, was "that he wishes to wait for fine weather in order to obtain the best results. This, to my mind, is to act less scientifically than you would."[85]

Biot explicitly drew ties between social position and intellectual work.

> It is very unfortunate for Science to see a man with such sagacity always considering the results from the artistic point of view and never at all for the noble intention of contributing to the progress of discovery in general, but I have given up preaching to him on the point. You, Sir, are in a very different social and intellectual position, I do not hesitate to beg you

[83] Ibid., 227.
[84] "Biot to Talbot, 27 May 1839," 1937–4840, NMFP&T Archives.
[85] "Biot to Talbot, 11 February 1840," 1937–4856, NMFP&T Archives.

urgently to make for our Gentlemen some of your beautiful pictures by the process that you have discovered and given to other people.[86]

The "artistic point of view" was equated with the venal world of profit and surface appearances. The noble intention of contributing to the progress of discovery could precisely be considered so because those involved retained their disinterest through aristocratic privilege.

Worker representation(s)

Arago, even more explicitly than Biot, tied the photograph to a specific vision of political practice. His speech before the Chamber of Deputies has assumed a near canonical place in the literature on early photography as a statement of the democratizing function of the daguerreotype.[87] His insistence that this new form of representation was within reach of all was and is still read as a claim that political representation was as well. This claim was not, as it is so often taken, a simple structural homology or expression of zeitgeist. Arago wielded the photograph directly in one of the most heated efforts of the July Monarchy to redraw the boundaries of the electorate. If his earlier epistemic claims were configured as essentially political, now his political claims became epistemic. The "judgment" required of electors was an act of perception, and Arago turned to the photograph to prove that even the disenfranchised knew what they were seeing.

At the same time that he was pleading for Daguerre's pension, Arago headed a movement by the Radical Party to bring the issue of universal manhood suffrage before the Chamber of Deputies. In 1839, Arago was one of the central members of the newly forming Radical party. That year, a broad-based opposition coalition had united to oust the conservative minister Molé. After its success, the coalition dissolved, and as Odilon Barrot's dynastic left and Thiers' center left pulled away, the republicans found themselves an increasingly unified party.[88] They took the name "Radicals" in the English style because the

---

[86] "Biot to Talbot, 7 March 1840," LA 40–31, Lacock Archives. The word "beautiful" was an interesting addition to this critique of the artistic point of view. Joel Snyder has suggested that Biot was engaged in a play on words. Talbot's term for the process, "calotypes," was taken from the Greek word καλόσ, meaning "beauty" or "good." Biot's pun would thus serve as yet another tie, binding these two men known for their work with ancient languages.

[87] The speech before the Chamber of Deputies is often confused with the speech before the Academy of Sciences, as the published texts of the two are the same. The first published account of the Daguerreotype process was in the *Comptes Rendus* for the week following Arago's announcement at the Académie des Sciences. Rather than give the text of the Academy speech, however, he printed his speech from the Chamber of Deputies, with additional notes, giving the technical details of the process. François Arago, "Rapport sur le daguerreotype," *CRAS*, 9 (1839): 250–267; Arago, *Rapport sur le daguerreotype* (Paris: Bachelier, 1839); Arago, *Rapport sur le daguerreotype, avec les textes annexes de C. Duchatel et L.-J. Gay-Lussac*, préface de Jean Bérezné (La Rochelle: Rumeur des ages, 1995), among others.

[88] Jardin and Tudesq, *Restoration and Reaction, 1815–1848*, Elborg Forster, trans. (Cambridge: Cambridge University Press, 1987), 120.

name "republicans" was at that point illegal.[89] But more than that, none of them wanted a repetition of 1789; only in 1848 did they abandon their efforts for a liberal monarchy. Until that point they wanted to include the proletariat in a way that would defuse their volatility without leaving them in charge.

The central platform of the radical deputies was electoral reform, founded on the conception that the most effective means of addressing the miserable condition of the working classes was to allow them a political voice. The push for wider suffrage began in earnest after the restructuring of 1839. In September, The Committee for Reform was formed in the Chamber of Deputies with Laffitte as the president and Dupont de l'Eure as the vice president.[90] Arago and Martin de Strasbourg shared the secretarial duties. They circulated petitions throughout the winter of 1839–1840 calling for

1. universal and direct suffrage;
2. electoral rights for the National Guard;
3. abolition of the political oath;
4. election in two steps;
5. fixing a minimum of 600 electors per college;
6. uniting the electors of a department into a single college; and
7. addition of the second jury list.

The first of these platforms, the demand for universal suffrage, was both the least likely to succeed and the most emblematic of the project. The radicals insisted precisely on its uncompromising claim that the public include every adult male.

As the committee prepared its platform, Arago continued to press the case of Daguerre. By June 15, 1839, Arago had negotiated a bill with the Minister of the Interior, who brought it before the Chamber of Deputies. The bill proposed that the Chamber purchase Daguerre's process in the name of the state to "bring it before the public."[91] There was no way, it pointed out, that the inventors could protect their work with a patent:

> As soon as it is known, everyone will be able to use it. The biggest mal-adroit will make pictures as accurately as an experienced artist. It is thus necessary that this process belongs to the entire world or else remains unknown. And what just regrets every friend of art and science would express, if such a secret should remain impenetrable to the public.[92]

The Chamber was to vote on July 3, 1839. Before they did, Arago gave the first full-scale report on the process of photography. He was still the only man

[89] Daumas, *Arago: La jeunesse de la science*, 217.

[90] "Réforme électorale," *Le National*, May 16, 1840.

[91] "Bill presented before the Chamber of Deputies, France, June 15, 1839," *Photography in Print: Writings from 1816 to the Present*, Vicki Goldberg, ed. (New York: Simon and Schuster, 1981).

[92] Ibid.

(besides Daguerre) to have witnessed the actual making of daguerreotype. He stressed to his audience the ease and straightforwardness of the operation:

> The daguerreotype does not involve a single manipulation that is not within reach of everybody. It does not assume any knowledge of drawing, it does not require any manual dexterity. By following, step by step, a small number of very simple prescriptions, there is no one who should not succeed as certainly and as well as M. Daguerre himself.[93]

Here indeed was the means to make every man a competent representer of the world around him. The price, Arago stressed, was no obstacle. It may seem, he admitted, that the use of silver plates, which cost three or four francs each, made this a more costly pastime than the use of paper. But one had to keep in mind that a single plate could be used successively for hundreds of pictures. The true cost of each picture was thus the cost of the ingredients, and this, Arago assured the Chamber, was negligible. Paper was cheaper. But with its "confusion of images" and "lack of certitude," it remained a "curious physics experiment," while silver plates guaranteed a resemblance so unmistakable that they brought representation to the hands of all.[94]

As he spoke, Arago passed around images made by Daguerre of well-known Parisian landmarks: Notre Dame, the Tuileries, and the Pont Neuf. As these "pass before your eyes," said Arago, "you can each imagine the immense use one could have made of such an exact and prompt means of reproduction during the Egypt expedition."[95] Certain facades that have been lost forever to the "monde savant" through vandalism could have been preserved. The daguerreotype would have been invaluable in copying hieroglyphs. Dozens of years and the work of "legions" could be replaced by a single man armed with Daguerre's apparatus. In France itself, Arago continued, the Commission of historical monuments was at work cataloging the patrimony of France. Photography was destined for a great role in this national enterprise, bringing with it that rare combination of economy and perfection of product. The Chamber, presumably in favor of both economy and perfection, voted unanimously to grant a pension to the inventors.

At the same time that Arago was describing to the Chamber his vision of monument-photographing multitudes, he was creating it from the personnel and materials of the Observatory. Noël Lerebours was the principal optician attached to the Paris Observatory. He made optical instruments, with a particular skill in lens making and arrangement. On Arago's suggestion, he turned his attention to making camera obscuras. Even as the fixing process itself remained a mystery, he constructed several devices suitable for *daguerreotypie*, and trained the young men of his workshop in their use. By the time

---

[93] Arago, "Rapport sur le Daguerreotype," 261. He reiterates the point that anyone can make a daguerreotype in *CRAS*, 10 (1840): 74.

[94] Ibid.

[95] Ibid., 251.

Daguerre's process became public knowledge on August 17, an entire team of workmen-photographers was ready for action.

By December 1839, Lerebours was able to display daguerreotypes from around the world in his shop on the Pont Neuf. By 1840, he was able to publish the first of many installments of his *Excursions Daguerriennes*, a collection of engravings made from daguerreotypes that came out in serial form over the space of several years.[96] The book sold wildly. Lerebours' images, the first example of what would be a flourishing field of photographic tourism, were the toast of Paris in the early days of *Daguerreotypemanie*.[97]

Arago was a steadfast supporter of the project. He passed around examples of the work at the session of the Academy of Sciences. *Le National*, the radical newspaper edited by François' brother Étienne Arago, gave the following report:

> M. Arago placed in front of the Academy several daguerreotype prints, which M. Lerebours has just received from Rome. These fine prints, representing the principal Roman monuments and some scenes of the Italian countryside, are remarkable for the vigor of their tints, even though they were taken in winter. The correspondent of M. Lerebours is a simple worker out of the optician's workshop, which can reassure those who claim that M. Daguerre's procedure can only succeed between the hands of a savant.

Here, then, was the realization of Arago's ideal. The same paper printed updates of the electoral reform petitions and reminded readers of their importance.

The push for electoral reform reached its denouement on May 16, when Arago presented the suffrage petitions to the Chamber of Deputies. At least two versions exist of the speech he presented before the Chamber. The first, a draft existing in Arago's papers, seems to correspond to the copy he had brought with him with the intention of reading. The second, an account published *le National*, includes the numerous interjections and distractions that kept Arago from sticking to the scripted text. His efforts to invoke the "simple artisan" of the workshop were countered trope for trope with the anxiety of the unruly masses.[98]

The petitions, Arago claimed, contained 240,000 signatures, more than the 200,000 or so men who made up the electoral body.[99] Other sources, counting only those names that appeared in *le National*, put the number of signatures at 188,000.[100] Arago's task in presenting them was to dispel the aura of danger

---

[96] Lerebours, *Excursions Daguerriennes* (Paris: Rittner et Goupil, 1840).

[97] Heinz Henisch and Bridget Ann Henisch, *The Photographic Experience, 1839–1914: Images and Attitudes* (University Park, PA: Penn State University Press, 1994), 86.

[98] It is not entirely clear which of the two versions was actually delivered. Another possibility is that *Le National* decided to edit out that particular paragraph in its published account. F. Arago, "Sur la réforme électorale," *Oeuvres complètes de François Arago*, J.A. Barral, ed. (Paris: Gide et J. Baudry, 1865), vol. 12.

[99] "Chambre des Députés," *Le National*, May 17, 1840, 1.

[100] Jardin and Tudesq, *Restoration and Reaction*, 126.

in which they were shrouded. He had to provide a way of ensuring his fellow deputies that admitting the working class into the political process would not lead to social anarchy.

The fundamental principle of the French government since 1830, Arago stated, was national sovereignty. And yet, he continued, how could anyone claim this principle was well served if only one man in 40 over the age of 25 held the right to vote? The very portion of the population without the vote, moreover, was that responsible for the lion's share of contributions to the state. It was both useful and natural, Arago declaimed over an agitated crowd, that the class of people deprived of their political rights should demand them.[101]

Opponents of universal manhood suffrage claimed that the laborer had no capacity for governing. Yet the only capacity needed, Arago responded, was the ability to distinguish an honest man from a dishonest one, and this skill could be found equally distributed among the classes of society. The people were completely capable of recognizing merit when they saw it. He turned to the example of the Convention.

Arago: The Convention was nominated by the generality of citizens, the Convention will serve to prove that the population, when called on to use its electoral right, is not exclusive, that it can nominate in all classes of society; that it can look for merit wherever it appears.

M. Tupinier: In the cabarets! (commotion)
[...President quiets chamber]

Arago: I set out to prove the people know how to find merit, and that they will choose it wherever they believe to perceive it. Well! Good God, Gentlemen, current electors also appoint according to appearances; they vote for whoever appears to them to have the most merit.

The capacity to participate in electoral politics came down to an act of perception: could the laborer recognize "merit" as well as the property owner? The interjection from the crowd points to the widespread anxiety that this perception may be led astray by the culture of spectacle and its privileging of surface appearances over deep truths. Arago's response was not to deny that the working classes either frequented cabarets or were swayed by appearance, but to point out that appearances were all anybody had access to.

Arago sought to further rally support for *les classes ouvrières* by citing the numerous ways in which they "radiated" the glory of France at home and abroad. The lightning rod, the loom, the steam engine, and the locomotive were visible monuments to artisanal skill. In the version of the speech preserved in his *Oeuvres*, Arago ended the list with the twin emblems of observatory work: the precision clock and the telescope. These devices were among the first rank of scientific invention. "Well, then, go study the history of optics" continued Arago, "and , there too, you will find simple artisans striving with remarkable skill...to vanquish the causes of irregularity that seem inherent in nature,

---

[101] *Le National*, May 17, 1840, 2.

in the very essence of the materials employed."[102] Arago thus configured the working classes as an ultimately stabilizing force, by the very virtue of their participation in depicting the world.

Chamber members were not left to guess who the worthy telescope maker might be. Only weeks before, Arago had arranged for the Observatory's optician, Lerebours, to make a daguerreotype of the inside of the Chamber of Deputies.[103] Although supposedly to honor the spot where Daguerre received his state pension, the event also took place shortly before Arago's presentation of the petitions on universal manhood suffrage. Thus when Arago praised the "simple artisan" of telescopic talent, he invoked a living figure who, so shortly before, had demonstrated in the flesh his facility at representation.

Or so he intended. The report of *le National* indicates that Arago's talk may not have gone as planned.[104] Relentless heckling cut him short after the lightning rod and the loom. Without ever reaching Lerebours and the workers of his shop, he rushed instead to the end, whereupon it was decided that the demands of the petitions would not be voted upon. The vote ended in a landslide failure for the reformers. Arago would have to wait until 1848 to implement his vision of universal manhood suffrage, and then only to find it all too fleeting an apparition.

Conclusion

In 1842, Biot published a follow-up article in the *Journal des savants* about the printing of the *Comptes Rendus*. He, by and large, acknowledged the service they had done by supplanting the role of uninformed science journalists. But he still cautioned that scientific results should not be presented briefly, stripped of the conditions in which they were produced. What better example of the dangers of publicity, he claimed, than photography, "the very studies which were currently exciting the most interest in the Academy of Sciences."[105] The excited public gawking at daguerreotypes saw only superficial resemblance and believed that they knew what they were looking at. Ignorant of the complexities of the mechanism involved, they took the image of the thing for the thing itself. Yet the true referents of photographs, Biot warned, were "objects of research which escape the senses and are only graspable by the discussion of complex facts." The distinction between visible light and photogenic rays, hard won by Biot in the enclosure of his dark room, was absolutely crucial for interpreting the relation of the photographic plate to the world. Omitting any discussion of this fact, to free the photograph from the conditions of its production,

---

[102] Arago, "Sur la réforme électorale," vol. 12, 612.

[103] A.V. Simcock, *Photography 150: Images from the First Generation* (Oxford: Museum of the History of Science, 1989), 10.

[104] *Le National*, May 17, 1840.

[105] Biot, "Comptes rendus hebdomadaires des séances de l'Académie des Sciences," *Journal des savants* (1842): 265–292; reprinted in Biot, *Mélanges scientifiques et littéraires*, vol. 2, 285.

was to destroy any ability to signify. "Nobody would be able to appreciate them, ... The results would have the air of miracles, ... It would be better not to speak of them at all than to announce them, mangled like that, in the *Comptes Rendus*."[106]

And thus for Biot himself, the photograph sat at the center of a web of materials, techniques, and cultural performance that bore directly on the question of how public science should be. The epistemic status of the photograph, its reference to some mysterious, invisible world, was inseparable from the paper it was printed on. The way that Biot made photographs, the careful attention to the chemical reactions that made up the process, formed the crux of Biot's argument about why his photographs could stand in for the world. Biot performed this work in the same darkrooms he had used for his work in rotary polarization. It involved, as had his polarization work, a mastery not only of optical techniques, but also of the messy world of test tubes and chemical baths. And, like his use of polarization to explore the boundaries of life and nonlife, it was an intensely private exploration of an undepictable realm, whose traces could only be understood by the initiated few.

Biot opposed his work to the "artistic point of view" of the inveterate showmen, Daguerre and Arago. He could not and did not try to separate the shiny silver plates from the culture of spectacle that embraced them. Arago as well invoked the crisp clean lines of the daguerreotype as the key to its ability to represent. The uncanny resemblance of the photograph, for Arago, mitigated the need to ground one's trust in an intimate familiarity with its production. Arago himself never made any daguerreotypes. And he treated those of his Observatory employees as unproblematic representations of the visible world. Both the production and appreciation of images became the easiest of all affairs.

Throughout the 1840s, Biot's constant nagging had little effect on the enthusiastic acceptance of the daguerreotype as faithful representation. It is worth noting, however, that the sharp lines of the daguerreotype's silver plates fell rapidly out of favor in the 1850s. Increasingly, the photographic image circulated as a piece of paper. And, as we shall see, this was not the only aspect of Arago's regime of transparency to run into trouble.

---

[106] Ibid., 286.

6

Illuminate All Eyes: Colonial Markets and the Problem of Freedom

As students at the École Polytechnique, Arago and Biot had learned that the study of color was a moral affair. Gaspard Monge, their instructor in Descriptive Geometry, had introduced the topic with the statement that, "in the judgment that we make on the colors of objects, there enters, so to speak, something moral."[1] Monge had leveled this claim as an attack on the Newtonian assumption that all colors could be analyzed physically. As Arago and Biot began their own scientific careers, they came to disagree profoundly on precisely that point. Was color perception a personal, immediate experience, or an analyzable one? The question continued to divide their work, even though neither of them continued Monge's somewhat quaint use of the word "moral" to describe it. Yet the word "moral" remained interlaced throughout their work, as they, and the rest of France, struggled to envision the foundations of a new society.

The meaning of the word "moral" changed in the course of the nineteenth century. The original definition of the French Academy, which remained constant from 1694 to 1798, was "relating to *moeurs*," where *moeurs* were in turn defined as "Natural or acquired habits for good or for ill in all that regards the conduct of life."[2] By using the word moral, Monge had thus been claiming that a person's habits and personal inclinations played a role in how they saw colors. But the question of *moeurs* quickly shaded into questions of community, social organization, and correct modes of behavior. By 1835, the dictionary of the French Academy added a new line to the heading of "moral," defining it as "having *moeurs*, having principles and a conduct conforming to morality."[3]

As France moved to found its political legitimacy on public opinion, a nervous anxiety hung over the question, "how does one ensure the morality of a people?" Responses fell roughly along two lines: regulation and regeneration.

---

[1] Monge, "Mémoire sur quelques phénomènes de la vision," *Annales de chimie* 3 (1789): 135.
[2] *Dictionnaire de l'Académie française*, 5th edn. (Paris: F. Didot frères, 1798), 115.
[3] One sees a new addition for the word "moral" in 1835, "Qui a des moeurs, qui a des principes et une conduite conformes à la morale," *Dictionnaire de l'Académie française*, 6th edn. (Paris: F. Didot frères, 1835), 2: 229.

The first, proposed by the forces of conservative tradition, equated morality with Christian principles and the teachings of the Church. The second, proposed by the advocates of the republic, held that only the removal of existing inequalities could give rise to a spiritually regenerated and moral people. The question of ensuring morality became all the more pointed as demands to expand the domain of the public grew through movements for working-class suffrage and the abolition of slavery.

This chapter covers Arago and Biot's participation in the anxiety of the 1830s and 1840s over questions of morality, freedom, and social organization. It focuses particularly on the issue that posed the greatest threat to France's own sense of morality in the nineteenth century: the colonial sugar industry and concomitant system of slavery. Their relationship to the institution of slavery was, in both cases, somewhat accidental. Biot only became involved in the sugar industry after he had begun to study the substance as part of his investigation into the optical activity of living matter. And Arago had the question of slavery largely dropped in his lap when he became Minister of the Marine and of the Colonies in 1848. But in both cases, their work was bound up with the larger question of how to ensure proper behavior among all classes of society.

Biot's vision, expressed in several essays on political economy, was along the lines of an essentially mercantilist economic organization that emphasized the wealth of the land, a centralized colonial arrangement, and the essential morality of keeping the workers under the control of this system. Key to this system was the need for strict regulation, keeping all participants behaving properly. In the 1830s, this regulation took concrete form when he proposed the use of the polarimeter as a way of standardizing the products of the sugar industry and tying together France's most extensive colonial enterprise. In the saccharimeter, the morality of the eighteenth and nineteenth century came together. The instrument depended on decisions about Monge's sense of the involvement of human will or volition. But it also had implications for the utopic sense of proper social arrangement and code of conduct.

Arago also thought about the organization of labor. He argued ceaselessly for greater autonomy for the working classes, emphasizing their ability to make their own choices. At stake were the definitions of liberty—were people free to act?—equality—did everyone experience the world the same way?—and fraternity—what implications did this have for the relations that bound people together? Nowhere were these issues more difficult to reconcile than within the slave colonies, particularly after 1789, when the colonies had officially become part of the mother nation. Arago's participation in the movement to abolish slavery, and the decree of abolition issued under him in 1848, tested the limits of the republican commitment to freedom.

Both Arago and Biot cast the issue of social organization as a moral one. As it became increasingly obvious that France was struggling with the problem of class difference, the question was how could one instill moral values equally in all segments of the French population. Biot's answer was that, as the differences were insurmountable, the only way to ensure proper behavior was through careful regulation. Arago, on the other hand, claimed that the differences were specious and the only way to ensure morality was to alleviate the economic gap that separated oppressor and oppressed.

Sugar and slavery in the colonies

Sugar, with its intriguingly vital nature, was an object of great theoretical interest to Biot. But it was also the product of an increasingly complex industrial process, and here, too, Biot played a role. His work was never completely separate from the constant demands for increased sugar production in the nineteenth century. Already in 1815, when, Biot first published his observations on sugar's effects on polarized light, France was in the middle of a sugar crisis.[4] The taste for sweetness had exploded in the eighteenth century, as the markets flooded with affordable sugar from the colonies.[5] Sugar was one of the key components of the triangular trade that moved slaves, sugar, and rum between Africa, North America, and the West Indies. The growth of French colonial holdings was largely driven by the sugar-producing slave plantations in the Caribbean.[6]

The Revolution of 1789 marked a blow to this precarious constellation of political, economic, and agricultural conditions.[7] In 1791 the slaves of Saint Domingue, the largest sugar producer, revolted.[8] In 1794, the National Convention abolished slavery and the slave trade in the French Empire. Although Napoleon reinstated both in 1802, the sugar crisis was further impacted by the disruption of sea trade brought about by the war with Britain, which made it difficult or impossible for France to import sugar from its colonies.

---

[4] *Bulletin des Sciences*, 1815; mentioned in Biot, "Comparaison du sucre et de la gomme arabique dans leur action sur la lumière polarisée," *Bulletin de la Société philomatique* (1816): 125–127.

[5] Sidney W. Mintz, *Sweetness and Power: The Place of Sugar in Modern History* (New York, NY: Viking, 1985); Moitt, Bernard, ed., *Sugar, Slavery, and Society: Perspectives on the Caribbean, India, the Mascarenes, and the United States* (Gainesville: University Press of Florida, 2004). B.W. Higman, "The Sugar Revolution," *Economic History Review*, 53 (2000): 213–236.

[6] The history of French sugar is closely bound to the history of French colonial involvement in the Caribbean. See, for example, Dale W. Tomich, *Slavery in the Circuit of Sugar: Martinique and the World Economy, 1830–1848* (Baltimore: Johns Hopkins University Press, 1990); Elborg Forster and Robert Forster, eds., *Sugar and Slavery, Family and Race. The Letters and Diary of Pierre Dessalles, Planter in Martinique, 1808–1856* (Baltimore: Johns Hopkins University Press, 1996); Verene A. Shepherd and Hilary McD Beckles, eds., *Caribbean Slavery in the Atlantic World* (Princeton, NJ: M. Weiner, 2000); Arthur L. Stinchcombe, *Sugar Island Slavery in the Age of Enlightenment: The Political Economy of the Caribbean* (Princeton, NJ : Princeton University Press, 1995).

[7] Robert Louis Stein, *The French Sugar Business in the Eighteenth Century* (Baton Rouge: Louisiana State University Press, 1988).

[8] C.L.R. James, *The Black Jacobins: Toussaint Louverture and the San Domingo Revolution* (1938; reprint, New York: Vintage, 1963); David Geggus, *Haitian Revolutionary Studies* (Bloomington: Indiana University Press, 2002); Laurent Dubois, *Avengers of the New World: The Story of the Haitian Revolution* (Cambridge, MA: Harvard University Press, 2004); Sibylle Fischer, *Modernity Disavowed: Haiti and the Cultures of Slavery in the Age of Revolution* (Jamaica: University of the West Indies Press, 2004).

In a desperate attempt to replace colonial sugar, Napoleon turned to the project of producing it from locally grown beets. He invested enormously in the effort. By 1814, there were 33 factories in France that had produced more than 35,000 tons of beet sugar.[9] It was in this context that Biot put in his word on the matter, claiming that both cane sugar and beet sugar rotated a polarized beam in the same direction and with the same intensity "which would necessarily be a new proof of their identity."[10] But the public remained suspicious of the alternative source.

Napoleon's crash program was ultimately short lived. With his fall in 1814, the beet sugar industry virtually disappeared. The Restoration government abandoned Napoleon's system of subsidies and threw its efforts into restoring the colonial economy. The treaty of May 30, 1814, restored to France a portion of its previous colonial holdings, the islands of Réunion, Martinique, Guadeloupe, and Guyana.[11] The focus on these plantation islands, known as the "major colonies," shifted more squarely to the production of sugar, and away from crops such as coffee and cocoa.[12] Sugar refineries were once again able to operate at previous levels. Although the famous "battle of the sugars" between beet and cane would loom large later in the century, in the 1820s, beet and cane planters worked together to ensure a protected market for themselves.

In the 1820s, France was reaching the apex of an upswing of colonial mercantilism. The Revolution and Napoleonic wars had dealt a heavy blow to France's colonial trade. After the Restoration, landowners, shipowner-merchants, and army men prevailed in their influence in Parliament, which enacted laws piecing back together the colonial economic system. Indeed, the reach of protectionism extended even further. Previously, only manufactured items had benefited from state protection. In the 1820s, as the government swung further to the right, producers of grain and livestock also began to call for the prohibition of foreign goods. One of the loudest voices in this debate came from a lobby of sugar planters. In 1822, Villèle, the ultraroyalist minister responsible for orchestrating the return of the émigrés (and himself a colonial planter from Réunion), coordinated a series of protective tariffs that almost entirely kept out foreign sugars.[13]

Under both the Restoration and July Monarchy, the sugar industry benefited from substantial state protection and an absence of taxes.[14] Both regimes returned to the policy of exclusion, where the colonies were allowed

---

[9] W.R. Aykroyd, *Sweet Malefactor* (London: Heinemann, 1967), 98.

[10] Biot, "Comparaison du sucre et de la gomme arabique dans leur action sur la lumière polarisée."

[11] Arthur Girault, *The Colonial Tariff Policy of France* (Oxford: The Clarendon Press, 1916), 51.

[12] Dale Tomich, *Through the Prism of Slavery: Labor, Capital, and World Economy* (Lanham: Rowman & Littlefield Publishers, 2004), 128.

[13] For Villèle's colonial roots, see Robin Blackburn, *The Overthrow of Colonial Slavery, 1776–1848* (London: Verso, 1988), 477.

[14] Roland Villeneuve, "Le financement de l'industrie sucrière en France, entre 1815 et 1850," *Revue d'histoire économique et sociale*, 38 (1960): 285–319.

to trade only with France.[15] An alliance of sugar planters and the extreme right pushed through an increase in the surtax of foreign sugar from 10 to 25 francs for 50 kg, effectively ending the importation of cheap sugar from India and Brazil. In 1829, France imported 73,769,000 kg of raw sugar from the colonies, compared to 229,000 kg from foreign sources.[16] The Antilles and Réunion devoted themselves entirely to sugar, and gave up what had been known as their secondary crops of coffee and cocoa.[17]

Despite substantial state protection, the sugar industry was a mess. In 1829, the Ministry of Commerce and Manufactures came out with a report, called the *Enquête sur les sucres*, on the sugar market that made it clear just how dependent colonial sugar was on the protective tariffs.[18] It also made it clear how expensive and inefficient the arrangement was.[19] Shortly after the appearance of the *Enquête*, Biot submitted his own recommendations to the Chamber of Commerce of Paris in an open letter to the director of the *Revue Britannique*, who was, incidentally, Saulnier, fils, the man responsible for bringing the zodiac of Denderah to France. The subject was not Biot's usual area of expertise, but he felt compelled to write, he said, because he was touched so deeply by the abuses of the present system and the "mortal wrong that they do to both agriculture and the morality of the people."[20] This work of agricultural policy was the only item he published in the twelve years from 1822 to 1834.

The letter, entitled, "On the provisioning of Paris," was a call to strong state direction. Paris relied on distant producers for its food. These producers were generally unable to bring their food to market themselves, so they relied on "parasite speculators" to buy their food and then sell it again. This, however, opened the door for abuses. Any salary these intermediate speculators received would only lower profits for the farmers and raise costs for the consumers. The Paris provisions market was, Biot claimed, overrun by such parasites. The administration needed to show a strong regulatory hand.

A free market, Biot claimed, was no good for foodstuffs. The economists' policy of laissez-faire laissez-passer might work for general goods, which could be completely abstracted from their origins. The particular living nature of agricultural products, however, required a more "careful observation of where the products come from in order to arrive at the consumer."[21] Other manufactured items could be rapidly multiplied according to the needs of commerce. Yet with agricultural products, the producer could not arbitrarily elevate the quantity of his products; he depended on the weather and the

---

[15] Girault, *The Colonial Tariff Policy of France*, 53.

[16] Ibid., 58.

[17] Ibid.

[18] Ministère du commerce et des manufactures, *Enquête sur les sucres* (Paris: Imprimerie royale, 1829).

[19] Émile Boizard, *Histoire de la Législation des Sucres (1664–1891)* (Paris: Bureaux de la Sucrerie Indigène et Coloniale, 1891), 19.

[20] Biot, *Lettres adressées au directeur de la Revue Britannique sur l'approvisionement de Paris* (Paris: Bachelier, 1835), 46.

[21] Ibid., 15.

seasons. The necessary role of food in sustaining life also rendered traditional economic models invalid. Usually, a severe lack of an item would result in an elevation in its price, which would in turn lower demand. Few people, Biot pointed out, would look to stock up on their linens during a cloth shortage. In the case of a famine, on the other hand, the first impulse was to begin hoarding food, an action only inflamed by spiraling prices. And the other end of the scale, overabundance, had harmful effects equally peculiar to the agricultural market. Where a typical speculator might keep his goods off the market to elevate prices, the perishable nature of food forced the agriculturist to sell as soon as he pulled the crop from the ground.

> These comparisons suffice to show that, if the commerce of other manufactured objects might be completely free and left to its own combinations, that which makes up the alimentary provisioning of the capital should be…very closely controlled (surveillé) with as much activity and constancy as enlightenment.[22]

Active and constant surveillance would thus save the agricultural market from the pernicious effects of speculation by ensuring that "the true price of each foodstuff is assigned to it."[23] Some, Biot, admitted, may be tempted to call this perfection of the market "a miracle." But, he pointed out, "I maintain that the miracle is the most straight forward and easy thing in the world."[24]

Biot's insistence on the particularity of the wealth of the continually producing land, as well as his conscious self-styling as a "*propriétaire*," echoed the stance of the eighteenth-century physiocrats, who sought to found citizenship rights on property ownership.[25] Yet there was a crucial difference. While the physiocrats demanded an absolute freedom of industry from the state, Biot defended the state's presence as the very thing that guaranteed the efficient working of the market.[26]

Biot ended his letter with an optical wish: "If only we could illuminate for all eyes what shines on our own with such vivid light!"[27] The proper role of the state, Biot claimed, was to keep open the paths of circulation between the provinces and capital and discourage the injustices that arose from excessive speculation. Biot thus urged the administration to take care in determining which of its agents were necessary and which were superfluous, and to ensure that the most direct route always remained open for trade. All this may seem self-evident, admitted Biot, and yet the "blindness" of the speculators was such

---

[22] Ibid., 20.

[23] Ibid., 19.

[24] Ibid.

[25] Pierre Rosanvallon, "Physiocrats," in *A Critical Dictionary of the French Revolution*, François Furet and Mona Ozouf, eds. (Cambridge, MA: Belknap Press, 1989), 763–769.

[26] The physiocrats have been discussed as forwarding an exemplary liberal ideology of private commercial relations occurring in a space outside of state control. As one moves into the nineteenth century, however, one sees a movement toward organized industry.

[27] Biot, *Lettres*, 14.

that they viewed the regulated trade of agricultural products as a violation of their rights.[28]

Biot had always emphasized that social economy was a moral problem. Removing the unnecessary steps would benefit agriculture, result in better produce in the Paris markets, and "inspire in the inferior classes of people, involved in the transactions of the public market, a sense of morality and self-respect."[29] Particularly important was the idea that inferior classes of people had to be kept in line and carefully regulated.

But in the sugar industry, as the 1829 *Enquete* pointed out, the kind of close surveillance Biot proposed was at odds with the system of self-contained plantations that had developed over the years.[30] Each plantation was in charge of its own milling and refining. In the first step, the cane was ground down into "sugar juice," made up of sugar and other foreign substances that prevented the sugar from crystallizing. The next step was the process of refining, where the impurities were removed and the crystallizable sugar skimmed off, leaving the uncrystallizable form as molasses. Refining was a bit of an art, with many different tricks available. The primary technique was cooking, which happened to be grossly inefficient. The *Enquête* estimated that, although sugar cane plants contained 18%–20% sugar, colonial plantations only had yields of 5%, losing roughly 8% in the grinding process and another 5% as molasses.[31] One great difficulty was that there was no single method of comparing the quality of sugar after refining.[32] The key issue, of course, was purity: how much of final product was crystallizable and how much was still in the form of molasses? But there was no direct test for purity, so inspectors relied instead on criteria such as the color, size, and firmness of the sugar crystals. The system was far from ideal: not only was color a rough stand in for purity, it was itself, as we have seen, a difficult means of standardization.

Biot had the perfect solution: the polarimeter. If the vital energy of sugar was responsible for rotating the plane of polarization, then the degree of rotation could be used to test just how much sugar there was in any given sample. In 1832, Biot fully laid out the use of the polarimeter to test the density and purity of end product sugar solutions. This memoir, the first on optics since his retirement to the countryside, was geared primarily toward the point of separating living and nonliving substances. But sugar played a large role in this division. He established a value α, which he called the force of molecular rotation. The force varied precisely, he showed, with the density of a sugar solution.[33] The desirable crystallizable sugar gave a strong polarimetric reading, whereas the uncrystallizable sugar did not.

---

[28] Ibid., 13.

[29] Ibid., 14.

[30] Dale W. Tomich, *Slavery in the Circuit of Sugar.*

[31] Ibid., 185.

[32] Ibid., 186.

[33] Biot, "Mémoire sur la polarisation circulaire et sur ses application à la chimie organique, lu à l'Académie Royale des Sciences le 5 novembre, 1832," *MAS*, 13: 1835.

By 1840, Biot was addressing his work directly to the colonial sugar fabricants.[34] He demonstrated various tests with the polarimeter to ensure them of its industrial applications. After establishing a density reading of 20.21%, he dried out the solution and found that it contained slightly more and 20% solid matter. This indicated that the sugar was entirely crystallizable. By comparing the polarimetric reading with the density of the solution, one could determine the relative proportions of crystallizable and noncrystallizable sugars. He directed his remarks, "to the colonials above all."[35] This technique could be a way for them to immediately test the quality of the sugar juice as soon as it was extracted. They could then compare localities, soil, and cultures. They could also test it as it went through the standard processes of evaporating, concentrating, and grinding to determine how these actions altered the percentage of crystallizable sugar.

Biot also offered his technique to the refiners. No longer would they be forced to rely on methods that were "highly vague, highly uncertain" to judge to quality of brute sugar.[36] The test could be done in fifteen minutes. The sellers as well as the buyers would know what they were selling or buying. There was much money to be made, Biot claimed, by substituting a "certain notion" for a "vague or mistaken" one.[37]

Biot's own participation was direct. Industrial researchers and sugar manufacturers would send samples of their product to him or the Academy of Sciences. He observed the rotation readings the samples gave when placed in a polarimeter and compared these results to the rotations of pure cane sugar that he had presented in his 1832 memoir.[38] Biot also provided a process for determining the portion of crystallizable cane sugar in a mixture of it and other sugars.[39] He made it clear that he intended his work to be of direct application to the sugar industry, by suggesting how it could be used to detect falsification, or to attack the problem of extracting usable sugar from molasses.

One problem that the state had to face in the regulation of the sugar industry was the practice of "falsifying" sugar. Shady merchants would cut their supply with a portion of glucose or starch sugar, or sometimes even chalk, plaster, or sand. One way to attack the problem was to try to find a particular chemical that would react with the false substance and form a precipitate that could be filtered out. But, the *Dictionary of Alterations and Falsifications of Alimentary*

---

[34] Biot, "Sur l'utilité que pourraient offrir les caractères optiques dans l'exploitation des sucreries et des raffineries," *CRAS*, 10 (1840): 264–266; also reported in "Académie des Sciences," *Le National* (February 19, 1840), 1.

[35] Ibid., 264.

[36] Ibid., 265.

[37] Ibid., 266.

[38] Biot, "Examen compartif du sucre de maïs et du sucre de betterave, soumis aux épreuves de la polarisation circulaire," *CRAS*, 2 (1836): 464–467.

[39] Biot, "Sur l'application des propriétés optiques à l'analyse quantitatives des mélanges liquides ou solides, dans lesquels le sucre de canne cristallisable est associé à des sucres incristallisables," *CRAS*, 16 (1843): 82–639.

Substances pointed out, there was a more general method, "due to M. Biot," which used the optical properties of sugar solutions.[40] Since the rotary power of sucrose was, as Biot had shown, different from that of glucose and starch sugar (and certainly different from that of sand), one could use the polarimeter to keep the honesty of the sugar sellers in line. In the 1840s, the municipality of Bordeaux, on a tip from the Chamber of Commerce, investigated some suspect merchants. Biot's close colleague, the chemist August Laurent, came in to analyze their product and found it contained up to 20% glucose.[41]

By the second-half of the century, Biot would be known as the father of saccharimetry, the science of sugar standardization. The polarimeter wound up being an absolutely crucial element in the manufacture and pricing of sugar. The price of a batch of sugar was determined directly by its polarimeter reading. The sugar industry, one of the strongest examples of organized industry in nineteenth century France, had the polarimeter at the center of its efforts at standardization and coordination.[42] The polarimeter was thus both the very mechanism by which sugar was assigned an exchange value that turned it into a commodity abstracted from its connection with the earth, and also the means by which one could establish its essential difference from the inert matter of the market. Biot had solved the problem of colonial trade: how to coordinate far-flung industries.

But distance was not the only characteristic distinguishing colonial industry. The sugar industry was a political lightning rod in the first-half of the nineteenth century, and never far from the question of slavery.

Competing moralities

The 1830s and 1840s were thus a moment of a revitalization of colonial sugar plantations, when France focused its gaze even more tightly on its remaining Caribbean colonies and slave systems. It was also at this time that the abolition movement began in earnest. The question had lain largely dormant until 1830, when the July Monarchy had ushered in, for the first time, a systemic political opposition. And the abolition of Colonial slavery became one of the planks of the opposition.

France had, of course, abolished slavery before. The first revolution had opened the experiment, freeing the slaves in all French colonies in 1794. But the disastrous consequences on the colonial sugar trade led Napoleon to reinstate both colonial slavery and the slave trade in 1802. France had officially

[40] M.A. Chevallier, *Dictionnaire des altérations et falsifications des substances alimentaires médicamenteuses et commerciales avec l'indication des moyens de les reconnaitre* (Paris: Béchet Jeune, 1852), 383.

[41] Ibid., 375.

[42] François Melard, *L'autorité des instruments dans la production du lien social: Le cas de l'analyse polarimétrique dans l'industrie sucrière belge*. PhD dissertation, École des Mines de Paris, 1996.

suppressed the slave trade in 1818, after pressure from the British. But they refused to allow for a mutual inspection of ships, and did little to enforce the law themselves. It was well known that, under the Restoration, a healthy clandestine trade continued unabated.[43] In 1833, a newly strengthened mutual search agreement between Britain and France ensured that the ban on the slave trade would be enforced. The fact that Britain outlawed slavery in its colonies in that year served as a further goad for the abolitionists.

Members of the opposition formed the *Société pour l'abolition de l'esclavage* in 1834, with the Duc de Broglie as its president and Odilon Barrot and Hippolyte Passy as its vice presidents.[44] Victor Schoelcher, a young habituée of salon life, was one of its most visible and tireless members. His father had recently sent him to the Americas on family business, and he returned with a lifelong dedication to end the slavery he saw in practice there. The work of the abolitionists gained some notice. In 1835, the word "abolitioniste" appeared in the supplement to the French Academy's dictionary dedicated to "new words born from the political storms."[45]

Arago joined his fellow opposition deputies in the abolitionist camp. He supported petitions by the *Société française pour l'abolition de l'esclavage*.[46] He was one of the first to sign an 1846 petition, calling for the immediate abolition of slavery.[47] His library was filled with books on the question, including ones by Henri Wallon, Pierre Paul Castelli, Friends of Freedom, and Moreau de Jonnes.[48] He worked with Victor Schoelcher, circulating petitions and organizing antislavery conferences.

But the abolitionists were a small minority in a political culture that still sought to avoid major disruptions in its colonial industry. Although public opinion was moving toward favoring an eventual abolition of slavery, there was still a great deal of anxiety about how it would happen. The most widely articulated fear was that immediate abolition would lead to chaos, as the slaves had not been properly prepared for their role as responsible citizens.

In 1837, the *Académie des sciences morales et politiques* proposed a contest on the historical conditions of the abolition of slavery in the West. Although historical in topic, the contest clearly spoke to current anxieties about how to introduce abolition slowly and peacefully. The Academy sought a solution that would prepare slaves for liberty while still guaranteeing a source of labor

---

[43] Tomich, *Slavery in the Circuit of Sugar: Martinique and the World Economy, 1830–1848* (Baltimore: Johns Hopkins University Press, 1990), 54; Serge Daget, "L'abolition de la traite des noirs en France de 1814 à 1831," Cahiers d'études africains II, no. I (1971): 14–58; Paul Michael Kiestra, *The Politics of Slave Trade Supression in Britain and France, 1814–1848: Diplomacy, Morality, Economics* (New York: St. Martin's Press, 2000).

[44] Tomich, *Slavery in the Circuit of Sugar: Martinique and the World Economy, 1830–1848* (Baltimore: Johns Hopkins University Press, 1990), 58.

[45] Nelly Schmidt, "Les abolitionnistes français de l'esclavage, 1820–1850 Une recherche en cours," *Revue Française d'Histoire d'Outre-Mer*, 1er semestre 2000, t. 87, no. 326–327.

[46] Horace Chauvet, *François Arago et son temps* (Perpignan: Edition des Anis de François Arago, 1954), 118.

[47] Lawrence C. Jennings, *French Anti-slavery: The Movement for the Abolition of Slavery in France, 1802–1848* (Cambridge: Cambridge University Press, 2000), 238.

[48] Horace Chauvet, *François Arago et son temps* (Perpignan: Edition des Anis de François Arago, 1954), 118.

for colonial plantation owners.[49] The winner was Édouard Biot. Biot *fils* had just finished a work on slavery as part of his father's project on ancient China. Slavery, he claimed, had been a natural part of the landscape of pre-Christian China, held in place by a racial hierarchy similar to the one in the colonies.[50] He then turned his attention to the West.

Abolition of slavery in Europe was, Biot claimed, a slow and gradual affair inseparable from the spread of Christianity.[51] When both master and slave shared the religion of Christ, they could not ignore its message of the equality of all men before God. This explanation was a convenient one for the Academy. It emphasized that abolition had not occurred overnight in the West with a simple proclamation. And by linking the process to the spread of Christianity, it provided a justification for the continuation of the practice in the colonies, where, it was reported in France, slaves remained largely outside the fold.[52] The Academy selected Biot as its gold medal winner, and awarded him the 1200 franc prize.

Édouard Biot's claim that abolition must be preceded by conversion to Christianity was echoed by the Conseil colonial as it scrambled for reasons to dissuade the Chamber of Deputies in Paris from voting for abolition.

> As you know, Sirs, some time ago we came to ask from you a rather large sum, in order to bring priests over to propagate religious instruction among the slaves. Well, then! You readily agreed to this request; and scarcely had these levites of the Lord arrived in the Colonies and begun their mission, than the Deputies of the Chamber, without waiting for the result, look to move prematurely on this large question of emancipation, even though they all know the consequences....[53]

Emancipation, the Conseil claimed, was part of a process that should only come after religious conversion. Although few in France attempted to defend slavery in principle, the general sentiment in France was that slaves should be freed only after some future process of amelioration.[54]

Jean-Baptiste Biot had himself proposed the necessary foundation of Christian religion in regulating an agricultural population. When he was in Scotland, he marveled at the parochial school system that elevated the moral and religious sentiments of even the lowest laborer. In his essay on the social economy of Scotland, he emphasized that this land, although not particularly

---

[49] Jennings, *French Anti-slavery*, 128.

[50] Édouard Biot, "Mémoire sur la condition des esclaves et des serviteurs gagés en Chine," *Journal asiatique* (1837) ser. 3 t. 3, 246–299.

[51] Édouard Biot, *De l'abolition de l'esclavage ancien en occident. Examen des Causes principales qui ont concouru à l'extinction de l'esclavage ancien dans l'Europe occidentale et de l'Époque à laquelle ce grand fait historique a été définitivement accompli* (Paris: Chez Jules Renouard, 1840), xi.

[52] *Le Moniteur universel*, February 16, 1838; cited in Jennings, 112.

[53] "Le compte rendu de la séance de 15 novembre 1839 du Conseil colonial," reproduced in Alain Grillon-Scheider, ed. *Canne, sucre, et rhum aux Antilles et Guyane françaises du XVIIe au XXe siècle* (Paris: Éditions du Ponant, 1987), 245.

[54] Jennings, *French Anti-slavery*, 101.

blessed by nature, was nonetheless one of the most productive and fruitful. The reason, he claimed, was in the sober, hard-working population itself, which had been formed from the system of parish schools. The key, according to Biot, was that each school was under the jurisdiction of presbyters. The schoolmaster of each parish first underwent an exam before the ecclesiastical court on religion and morals, as the standard school subjects. These schools served the entire population of Scotland, and, added Biot, "it is them, more than anything else, which gives [the people] the thoughtful, serious, regulated, moral and religious characters for which the Scots of middle and inferior rank distinguish themselves so eminently."[55]

The abolitionists were less convinced of the necessary step of religious conversion. Arago's older brother and anti-slavery activist, Jacques, referred to Christianity as "a shield more sure" than jail or flagellations for keeping slavery in place.[56] After the Academy's contest, the *Société française pour l'abolition de l'esclavage* shot back with a competition of their own for the best essay supporting the immediate abolition of slavery. They had less luck, however, and failed to come up with a single winner.[57] But the movement for immediate abolition was picking up. Hippolyte Passy first raised the question of immediate abolition on the floor of the Chamber of Deputies in 1839.[58] In 1841, Victor Schoelcher returned to France from a year-long voyage in the Southern United States and Caribbean Islands to document the conditions of slavery. He began to call widely for an immediate end to the institution of slavery, exhibiting whips and chains as the physical evidence of the less-than-utopian conditions he witnessed.[59] He became the strongest voice for "immediatism," which he compared to the progressive abolition undertaken by Britain. In 1843, he founded the journal *La Réforme*, along with Arago's younger brother Étienne, Louis Blanc, and other members of the left. Conceived primarily as an instrument for the demand of workers' rights, *La Réforme* served also as the primary venue for abolitionist writings in France.

The basis of the Opposition abolitionist argument was that slavery was incompatible with the values of freedom and equality ushered in with the Revolution. Who, asked Tocqueville in his series of public articles, *The Emancipation of Slaves*, had "illuminated" Europe with the love of freedom that made its people cry out against servitude?[60] We did, he answered, the

[55] Jean-Baptiste Biot, "Sur le mode d'éducation du peuple en Écosse et particulièrement sur un genre d'éducation très-influent appelé Écoles paroissiales," *Mélanges scientifiques et littéraires*, 228.

[56] Jacques Arago, *Souvenirs d'un aveugle, voyage autour du monde* (Paris: Hortet et Ozanne, 1839–1840), 39.

[57] Jennings, *French Anti-slavery*, 128

[58] Phillippe Vigier, "The Reconstruction of the French Abolition Movement under the July Monarchy," *The Abolitions of Slavery: From L.F. Sonthonax to Victor Schoelcher, 1793, 1794, 1848*, Marcel Dorigny, ed. (Paris: UNESCO Publishing, 2003).

[59] Nelly Schmidt, *Victor Schoelcher en son temps: images et témoignages* (Paris: Maisonneuve et Larose, 1998), 27.

[60] Alexis de Tocqueville, "The Emancipation of Slaves," *Writings on Empire and Slavery*, Jennifer Pitts, ed., trans. (Baltimore: Johns Hopkins University Press, 2001), 207; Tocqueville,

French. Even Christianity, despite its admonitions of loving one's neighbor, had resigned itself to slavery. Only with the Republic had the French given a practical meaning to the idea that all men were born equal. Moreover, Tocqueville emphasized, only the implementation of equality could ensure moral progress in the colonies. The colonists predicted a bloody anarchy if the slaves were freed. But, Tocqueville reported, when the English abolished slavery in 1833, they observed precisely the opposite. "With *les lumières* and the regularity of *moeurs*," came a desire for civilization.[61] With the abolition of slavery, the negro would acquire "certain rights which he had never enjoyed until now, and without which there could be no progress in morals or civilization."[62]

The liberal opposition's campaign to end slavery coincided with the efforts at electoral reform discussed in Chapter 5. When *Le National* referred to electoral reform as the "emancipation" of France, it tied together the problems of workers' rights and colonial slavery.[63] The denial of freedom, and the exploitation of one group by another, formed the central impediment to a properly functioning state. Arago, in his 1840 speech before the Chamber of Deputies, did not shrink from casting the issue as a moral one: "There is a real evil, a cruel evil, for which we must find a remedy."[64] The answer, for Arago, was to establish true equality among the citizens of France, by organizing the working class and giving them full voting rights. *Le National*, reporting his speech, foresaw the consequences of accepting his proposals:

> When the root causes of evil have disappeared, evil itself will soon disappear as well. Misery, daughter of monopoly and the oppression exercised by capital on labor, will diminish, and with misery, vice, and with vice, crime. Then a people will rise up that is more vigorous, more moral, and more intelligent.[65]

There was a strong moral element to the criticisms of the opposition: with the suffrage limited to 200,000, the government did not represent France. Biot had claimed that careful surveillance from above would inspire the inferior classes to morality. *Le National*, glossing Arago's work, claimed that only the eradication of economic inequality and misery would do that.

Both Arago and Biot were very concerned with "morality," or the proper conduct of behavior. But both had very different visions of the kind of social organization that would bring it about. Indeed, they had different visions of the kind of social organization that was possible. Arago was articulating a theory of self-governance that rested on the assumption that human beings, acting as free agents, could come together and make decisions on their own free

---

"De l'émancipation des esclaves," *Oeuvres complètes d'Alexis de Tocqueville*, Gustave de Beaumont, ed. (Paris: Michel Lévy frères, 1866), vol. 9, 275.

[61] Tocqueville, "De l'émancipation des esclaves," 281.

[62] Ibid., 289.

[63] "France, Paris, 16 mai," *Le National*, May 17, 1840, 1.

[64] Chambre des Députés, Séance du 16 mail," *Le National*, May 17, 1840, 2. Italics in original.

[65] "France, Paris, 20 mai," *Le National*, May 21, 1840, 1.

will. Biot, for whom free will was tempered by irreducibility of the unknown, required a higher power. Where Biot advocated a strict regulation of the inferior classes by an external authority, Arago advocated self-regulation and expansion of workers' rights.

"Is man free?"

The colonial enterprise posed certain problems for the advocates of the republican values of liberty, equality, and fraternity. Few members of the liberal opposition wanted to abandon the colonies. Indeed they usually advocated tighter integration into the French Empire. The abolitionist movement saw itself as extending the values of liberty, equality, and fraternity to all corners of the Empire. Yet this very act of extension pointed to the unequal positions of metropole and colony.

It was over a matter of colonial policy that the opposition began its break with the July Monarchy. At issue was the "Tahiti Affair." The admiral Dupetit-Thouars was at work in the South Seas consolidating French colonial holdings. In 1843, he annexed Tahiti and in the process removed, with some violence, an English missionary who had been on the island. Here was a conflict between the two modes of moralization. Was the proper path for Tahiti with the missionary, and Christianization, or with the French admiral, and integration into the French polity? Many were outraged at the mistreatment of a missionary. England demanded reparation, and the minister Guizot quickly gave it. The opposition leapt on the issue as a means of criticizing Guizot. *Le National* whipped public opinion into a fury over the supposed humiliation.[66] Arago's election address of 1846 revolved around Guizot's leniency toward the English missionary and his servile attitude toward the English.[67] Although Arago won his seat, the parliamentary elections of 1846 were in general a sweeping success for Guizot and deputies loyal to the government, giving the party of Louis-Philippe its first coherent majority.[68] With their voice eclipsed, the opposition deputies, Arago included, began to meet every day at the house of Garnier-Pagès. The administration soon clamped down on these meetings, which were outlawed by the *l'interdiction du droit de réunion*, intended to discourage plots to overthrow the government.

The *Campagne des banquets* that ran through 1847, and ended in the revolution of 1848, began as an attempt to get around the *l'interdiction du droit de réunion*. Since each attendee was supposedly paying for his meal, the banquets could not be classified as political meetings. In the first banquet, held in Château-Rouge, a public garden near Montmartre, the opposition deputy Odilon Barrot spoke out against the "moral disorder which menaces our society," and

[66] H.A.C. Collingham, *The July Monarchy: A Political History of France 1830–1848* (London: Longman, 1988), 321.

[67] Archives Nationales, 220.AP.18; Ibid., 322.

[68] André Jardin and André-Jean Tudesq, *Restoration and Reaction, 1815–1848*, trans. Elborg Forster (Cambridge: Cambridge University Press, 1988), 140.

pointing to the "evil in our *moeurs publiques*."[69] Beginning on July 18, 1847, 70 banquets were organized around France, with over 22,000 subscribers. Arago presided over banquets in Tours, Blois, and Perpignan.[70] His brother, Étienne, presided in Dijon, announcing that with the eradication of poverty and inequality, the science, arts and literature of France would flourish.[71] The officers of the National Guard organized a banquet in the XII arrondissement in Paris (the Observatoire-Panthéon district where Arago had served as a deputy) scheduled for February 19, 1848. They had first asked Arago to preside, but he declined, so they turned instead to the chemist Boissel.[72] The Ministry of the Interior had lost its tolerance for the banquets, and banned the February 19 event from taking place. Tension mounted as the day approached. Some deputies were willing to cede to the interdiction. But the more radical, Arago among them, stated their intention to go. On the day of the banquet, angry crowds and armed troops filled the streets. The opposition gathered at the offices of *le National*, to watch as barricades sprang up, firearms were looted from stores, and Paris erupted in yet another revolution.

The rioting, known as the *Journées de Février*, lasted three days and ended with Louis Philippe's abdication. The provisional government that took over consisted largely of former radical members of the Chamber of Deputies, Arago among them. They proclaimed the dawn of the Universal Republic, which extended rights equally to all male citizens. As the workers' government, they quickly passed a succession of laws which guaranteed the right to work, limited the working day, and granted universal male suffrage (Figure 6.1).

Arago took over the duties of Minister of the Marine and of the Colonies. One of his first actions was to send out officers of the Marine and Army to the colonies to ensure the loyalty of the French fleet.[73] The Universal Republic had no intention of relinquishing its colonies. The constitution of 1848 tightened the ties, binding the French Empire by declaring the colonies "integral parts of the Republic and subject to the same constitutional law."[74] A decree, most likely from the Minister of the Marine, called for the inclusion of the colonies in the Chamber of Deputies.[75] Arago had thus redefined the meaning of national representation (Figure 6.2).

As Minister of the Colonies, Arago found that the slavery question was now squarely in his lap. His abstract support of abolition faced the reality test, and he began to recognize some of the possible dangers. The threat of civil war

[69] *Le Constitutionnel*, July 12, 1847, and *Le Siècle*, July 12, 1847, 3. Quoted in William Fortescue, "Morality and Monarchy: Corruption of the Fall of the Regime of Louis-Phillippe in 1848," *French History*, vol. 16, no. 1, 83–100, 89.

[70] Élias Regnault, *Histoire de huit ans 1840–1848* (Paris: Pagnerre, 1851), 183.

[71] Frederick Engels, "Reform Movement in France. Banquet of Dijon," *Northern Star,* Decemeber 18, 1847. MECW Volume 6, 397.

[72] Collingham, *The July Monarchy*, 403.

[73] Louis-Antoine Garnier-Pagès, *Histoire de la Révolution de 1848: 1 Avénement du Gouvernement provisoire*, t. II (Paris: Pagnerre, 1866), 2: 10.

[74] Horace Chauvet, *François Arago et son temps* (Perpignan: Edition des Anis de François Arago, 1954), 122.

[75] Ibid.

Fig. 6.1 The Provisional Government of the Universal Republic making its way to the National Assembly. Arago is in the center of the second row (second from left). (From *République française* (Paris: Imprimerie nationale, 1848).)

Fig. 6.2 Arago, Minister of the Marine and Colonies, astride the world. (From Louis Reybaud, *Jérôme Paturot à la recherche de la meilleure des républiques* (Paris: Michel-Lévy frères, 1849).)

in the colonies loomed large, and fellow members of the provisional government raised the possibility of uprisings against the white population and losing naval position to the English.[76] A delegation of colonial planters visited Arago to plead their case. On February 26, Arago gave a speech before the colonial governors, calling for patience and indicating that the decision would not be hurried.[77] Soon after, he sent out a circular announcing that the question would be decided by the National Assembly, whenever it was formed. Louis Blanc accused him of giving in to pressure from the colonists.

But Arago was not completely done with the issue. As the lobbying colonists were weakening his resolve, he had written the fellow abolitionist Schoelcher, "Come. I need you."[78] Schoelcher, who had been in Senegal, returned to France as soon as he heard news of the revolution. He arrived on March 3, to find Arago's message waiting for him. He rushed to find Arago and make the case for immediate action. He "spoke to him from his heart," "pressed him…begged him."[79] He posed the question: "Is man free?"[80] Persuaded by Schoelcher, Arago took up the question of abolition himself. He drafted a decree appointing a commission "to prepare with the smallest delay the act of immediate emancipation in all of the colonies of the Republic."[81] The commission did act fast. By April 23, they presented the declaration of abolition. The decree began with an opening clause laying out the principles of the issue:

> The provisional government,
>
> Considering that slavery is an assault against human dignity;
>
> That by destroying the free will of man, it suppresses the natural principle of right and duty;
>
> That it is a flagrant violation of the republican dogma: Liberty, Equality, Fraternity.[82]

The decree cast abolition as a universal extension of the rights of the first revolution. The primary complaint against slavery was its destruction of free will. The rights and duties of a republic, based on universal suffrage, after all, were predicated on the ability to act according to one's will. Members of the provisional government described the revolution as operating on the principle of "the emancipation of humanity."[83] For the Universal Republic to live up to its claims, it could not restrict the category of humanity to a single race.

---

[76] Ibid., 119.

[77] Nelly Schmidt, *Abolitionnistes de l'esclavage et réformateurs des colonies, 1820–1851, Analyse et documents* (Paris: Éditions Karthala, 2000).

[78] François Sarda, *Les Arago: François et les autres* (Paris: Tallandier, 2002), 285.

[79] Ibid.

[80] Chauvet, 116, Ibid., 283.

[81] Sarda, 285; *Le Moniteur*, March 5, 1848.

[82] "Décret d'abolition," *D'une abolition, l'autre: Anthologie raisonnée de textes consarcrés à la seconde abolition de l'esclavage dans les colonies françaises*, Myriam Cottias, ed. (Paris: Agone, 1998), 17.

[83] Garnier-Pagès, *Histoire de la Révolution de 1848*, 1.

Arago's political star would rise even higher in the following month. The provisional government had moved quickly to provide its republic with elections. They decreed that France would be divided into as many electoral sections as there were representatives to elect. They gave Arago's brother-in-law, Mathieu, the task of calculating the divisions.[84] The newly enfranchised of France voted for a National Assembly in April, and the Assembly then voted for an Executive Committee. Arago, who received the most votes in this second election, served as its head. As Arago's closest political ally, Lamartine, remembered it,

> The name of Arago was saluted by unanimous acclamations. He possessed the twofold charm which so fascinates an intelligent people; science, a species of right divine which never meets an obstacle in France; and the reputation of honesty, to which every head bows with reverence.[85]

Arago's charm thus rested on two elements: the universality of science and his reputation for transparency.

At the peak of official power, Arago found the balancing act a delicate one. With the other members of the Executive Committee, he sought both to include the uneducated masses that constituted the real productive force of France, and to render anodyne the potentially destructive influence that their ignorance and superstition may present. He assumed the authority to speak for the people while guaranteeing to keep them under control. Yet Arago's regime of transparency, which rested on an easy communication among equals, had trouble accounting for the fault lines produced by social differentiation. Already his position as Minister of the Colonies pointed to the contradiction that would lead to his downfall. Here, the conflicting requirements of transparency and universality came to a head. Universality required that the rules for France apply to the colonies. Transparency required that there be no central organizing authority. But, as Minister of the Colonies, Arago was that central organizing authority.

The source of difference that eventually sunk the Republic of 1848 was the split between the working and propertied classes. The political inclusivity he demanded could not absorb issues of class conflict. Even Arago's admirers admitted that his claim to represent the working class showed a lack of awareness of the differences that separated them. George Sand, who had substantial personal contact with Arago, expressed her own reservation over this blind spot.

> As good, as fine, as great as a man may be, from the moment he is born into either the nobility or the bourgeoisie, and develops there, he no longer understands the people. Arago, Lamennais, Béranger, Lamartine, yes certainly, all glorious men, great geniuses, great and fine characters!

---

[84] Ibid., 3.

[85] Alphonse de Lamartine, *History of the French Revolution of 1848*, Francis A. Durivage and William S. Chase, trans. (London: George Bell & Sons, 1888), 204.

and yet the predication of equality is in their eyes a mad and dangerous utopia. They love the people and honor them as much as they can but they do not believe in them one bit, they do not know them and they do not understand them.[86]

Karl Marx expressed similarly mixed feelings. He had spoken favorably of Arago in the 1840s, pointing to the ways that Arago's astronomical practice used the organization of labor to achieve more results than possible relying on unique individuals.[87] But his take on Arago and the other members of the provisional government was less positive. He found their claims to represent the working class ludicrous. The revolution of 1848, he claimed, was the final solidification of power by the bourgeoisie. Whatever their stated intentions, Marx had little hope for Arago and partisans of *"le stupide National"* as he called it.[88]

Marx attributed the failure of the Second Republic to a dialectic leading to the end of the bourgeois monopoly on publicity.[89] As the bourgeois strata created a space within the political realm for public opinion, it configured this opinion as universal rather than tied to class interest. The consequent expansion of political inclusion to wider and wider definitions of the public meant that nonbourgeois elements came into possession of "the weapons of publicity forged by the bourgeoisie."[90] At the heart of this process lay the issue of electoral reform. The liberal bourgeois reformers, like Arago, argued for universal suffrage as the natural consequence of their enlightenment views, only to find these views overturned by a group who associated them with narrow class interests.

Arago claimed to speak for the people, although he made few efforts to reproduce the sort of things the people were saying. This paradox at the center of representation toppled Arago from his lofty post. His position at the head of the Second French Republic would last less than four weeks. Cracks had already begun to appear over the abolition of private property, one demand of the working class that the Executive Committee refused to consider. Dissatisfied over the lack of support for the right to work, the workers of Paris took to the barricades again on June 23. The Executive Committee dispatched forces of the *guarde mobile*. Yet Arago, arriving with the troops at the barricade on the Rue Soufflot, tried to speak with them one last time.[91] He stood alone in the vast empty space before the barricade and demanded to know why the workers were not doing what they were told. This government, after all,

---

[86] "George Sand to Charles Poncy, 26 January 1844," in George Sand, *Correspondance*, 6: 410.

[87] Marx and Engels, "The German Ideology," in *Karl Marx: Selected Writings*, David McLellan, ed. (Oxford: Oxford University Press, 1977), 189.

[88] Marx and Engels, "Le milliard," *La Nouvelle Gazette Rhénane*, n. 247, March 16, 1849.

[89] Marx, "Critique of Hegel's Doctrine of State," *Early Writings*, ed. Quentin Hoare, R. Livingstone and G. Benton, trans. (New York: Random House, 1975), 57–198.

[90] Jürgen Habermas, *The Structural Transformation of the Public Sphere*, Thomas Burger, trans. (Cambridge, MA: MIT Press, 1991), 126.

[91] Maurice Daumas, *Arago 1786–1853, La jeunesse de la science* (Paris: Belin, 1987) 271.

had been created for them. Arago enumerated the list of things the National Assembly had done on behalf of the worker. They were taking up arms, he pointed out, against the representatives of their own interests. The reply from the barricades was that, while Arago was not a bad man, he had never been hungry, and in fact had no idea what it was to be a worker. A gunshot rang past his head. He left the barricades, and the troops moved in. Three days of fighting ensued, at the end of which Arago, like the rest of the Executive Committee, was forced to resign his position, having failed in both his goals of making the people embrace him as their own, and keeping them from destroying Paris. The democratic socialists, or democ-socs, held power until June, when the political spectrum swung back to the right and the forces of reaction came to power. On December 20, 1848, the Constituent Assembly proclaimed Louis Napoléon president of France, who immediately restored a royalist administration.

Biot's last laugh

Biot's experience of 1848 was very different from Arago's. While Arago oversaw documents condemning "the destruction of free will," Biot preferred to view the revolution as an example of just how little control one had over the events of one's life. He referred to it as

> the time of these public commotions, when one has no other duty towards them than to endure the events, without exercising any influence over them, as in the great afflictions of the soul from which one cannot relieve oneself.[92]

At just the moment when Arago had his greatest political impact, Biot resigned himself to enduring events over which he exercised no influence. The question of free will and freedom was at the center of the revolution of 1848. Was a human being free to act as he or she chose? Arago sided with the claim that yes, an assault against free will was an assault against his most cherished principles. But Biot had never given such credence to free will in the first place. Mankind was the subject of external, often unknown, forces.

The idea of freedom rested upon assumptions about the nature of transparency. A free choice could only be predicated on full knowledge of the situation. Biot's rejection of freedom was based on his belief in mysterious influences impossible to see. Arago's sense of freedom boiled down to the idea that everyone else had just as much access to knowledge as he did, and was therefore considering the same pieces of information when making decisions, and thus would make decisions in exactly the same way. Biot maintained all along that everyone did not have equal access to the world. Some could see more clearly than others.

---

[92] Biot, *Correspondance du chevalier Isaac Newton et du professeur Cotes, avec des lettres de plusieurs autres personnages éminents; Extraits du Journal des savants (cahiers de Mars, Avril, Mai et Juin)* (Paris: Imprimerie Nationale, 1852), 1.

But as the political pendulum swung to the right, Biot found himself once again in favor. Biot flourished under Louis Napoléon. Within weeks, the department of public instruction named him an officer of the Legion of Honor.[93] He had been nominated before, he reported, twenty-seven years ago. But the regime of Louis XVIII had been little inclined to bestow him with ministerial favors. No matter, he wrote in 1849, "to tell the truth, I do not regret it a bit. For, in most of the circumstances which we have traversed, [ministerial favors] would have caused me more embarrassment than pleasure."[94] Biot also saw his vision of polarimetric sugar surveillance was written into law under the administration of Louis Napoléon. In 1851 the government passed a law called the "Loi substituant la saccharimétrie aux types," which elevated saccharimetry to the standard used to judge sugar quality.[95]

Arago fared less well under Louis Napoléon. Their quarrel came to a head as history repeated itself as farce, and the president declared himself the Emperor Napoléon III on December 2, 1852. On his ascension, Napoléon III reinstated the loyalty oath that had been abolished in 1848. Arago refused, with the consequence that he would be stripped of all official positions, most notably the directorship of the Observatory. Victor Hugo, outraged, derided the act in his *Napoléon le Petit*. "Free astronomy," he wrote, "is almost as dangerous as the free press. Who knows what transpires in the nocturnal tête-à-têtes between Arago and Jupiter? If it was M. Leverrier, alright, but a member of the provisional government!"[96] Arago prepared to accept his dismissal. It was with a heavy heart, he wrote in his resignation, that he thought of the optical instruments he had worked so hard to collect fall into "enemy and malevolent hands."[97]

There is little doubt to whom these enemy hands belonged. Biot and Guizot had been hoping for awhile to rid the Observatory of its Cromwell. Even before the revolution, Guizot had masterminded a plan to oust Arago from his position by cobbling together the votes and presenting Arago's own student, Urbain Leverrier, as his rival.[98] The Revolution had thwarted his plan, but already in 1849, Louis Napoléon's administration was chipping away at Arago's reign. The Ministry of Public Instruction reduced the number of assistant astronomers from six to four. Arago set out to complain, but was unable to find an audience with the Minister.[99] The conflict over the loyalty oath seemed the final move to rid the Observatory of his influence.

But Arago managed to hang on once more. Arago's resignation letter also revealed his intention to alert all the foreign scientific academies of his

---

[93] Dumas proposes him for Legion of Honor, "Dumas au Ministre de l'Instruction Publique et des Cultes," January 29, 1849, Archives Nationales, F17 40053.

[94] "Biot à Monsieur le Ministre secrétaire d'Etat, au département de l'instruction publique," Paris le 4 mai 1849. Archives Nationales F17 40053.

[95] Boizard, *Histoire de la Législation des Sucres*, 383.

[96] Victor Hugo, *Napoléon le Petit* (London: W. Jeffs, 1863), 206.

[97] Maurice Daumas, *Arago 1786–1853, la jeunesse de la science* (Paris: Belin, 1987), 276.

[98] Ibid., 238.

[99] "Beautemps-Beaupré à Monsieur le Ministre," December 5, 1849. AN F17 13569.

situation, as well as his numerous colleagues abroad. Napoléon III, sensing a backlash, arranged for an exception. Arago stayed on, and the fight over the Observatory would have to wait until after his death. Arago had already begun preparing for this moment, carefully vetting his brother-in-law Mathieu (who shared his radical politics) for the position of Observatory director. Within two weeks of his death in 1853, on October 17, 1853, the Bureau of Longitudes held a convocation to decide who the next director of the Observatory would be. They appointed Mathieu.

Biot, claiming to be so profoundly afflicted that he could not sleep at all that night, immediately began planning a way to wrest control of the Observatory away from the Arago camp. The next morning he contacted Leverrier, and convinced him to petition the Ministry of Public Instruction to appoint a commission in charge of naming the director. Leverrier asked that the commission be kept entirely secret, and be headed by Biot.[100] That same day, the Ministry sent a letter to all the members of the Bureau des Longitudes voiding their previous night's decision, and assuming control of the appointment process. The Bureau immediately protested that usurping their decision was downright insulting.[101] But the Ministry persisted, and soon appointed a committee of Biot, Leverrier, and Charles Baudin. Admiral Baudin, the president of the Bureau of Longitudes, disapproved of the whole process, but Biot had a sure ally in Leverrier. This young astronomer had started his career as a bright protégé of Arago, who had encouraged and publicized his discovery of Neptune in 1846. Their relationship soured, however, when Leverrier went over Arago's head to propose to Louis-Phillipe that he take over direction of the Observatory. He was thrown out of the Observatory in 1847, and spent the next six years waiting for Arago to die. Few could have been surprised when Biot's commission reported back appointing Leverrier the director of the Observatory, as well as removing it from the aegis of the Bureau of Longitudes.

---

[100] "Leverrier au Ministre de l'Instruction Publique et des Cultes 18 October 1853", AN F17 13569.

[101] Charles Baudin, Le président du Bureau des Longitudes, "Note pour Son Excellence le Ministre de l'Instruction Publique," October 21, 1853, AN F17 13569.

Conclusion

—◦◦◦—

The last scene between François Arago and Jean-Baptiste Biot was meant to be a deathbed reconciliation.[1] Arago had spent the summer of 1853 in deteriorating health, and, by September, word had gone out that he did not have long to live. Biot made his way to the Paris Observatory to say goodbye to the man who had been the most constant presence of his professional life. He had known Arago for nearly five decades by this point, virtually all of it spent fighting with him. The argument between them had begun with optics and ranged over nearly every aspect of French culture. The conflict had defined their careers, just as they had defined themselves through their conflict. The sharpness of their criticisms only served to carve out more clearly their respective visions for science in the nineteenth century.

Biot had spent much of the previous fifty years in the shadow of his rival's brighter star. By 1853, however, Arago was bedridden and defeated. His eyesight had failed rapidly over the previous years, leaving him too blind to read or write. As if blindness were not an ironic enough demise for this champion of universal visibility, the root cause was pronounced to be diabetes, a disease whose treatment Biot was in the process of revolutionizing. As early as the 1840s, Biot had suggested the use of the saccharimeter in monitoring the urine of diabetics, but it was only after Arago's death that the practice became widespread.[2]

On that September day in 1853, however, Biot had other matters to discuss. He came to urge Arago to accept his salary as secretary of the Academy of Sciences, which had gone uncollected for months. Arago had refused to accept pay for work he had not done, and, being confined to bed, he had not made it to a session for some time, let alone found time to read the hundreds of letters or dozens of submissions he routinely received. However, Biot stressed that the Academy was not bothered with the lack of his physical presence. It wanted him to have the money and had sent Biot to ensure he got it.

Although the gesture was no doubt one of kindness, it touched on yet another of the issues that divided Arago and Biot: the practice of *cumul*, wherein French academics drew salaries for multiple positions without necessarily

---

[1] J.A. Barral, *François Arago* (Paris: Dusacq, 1853), 18.

[2] Biot, Jean-Baptiste, *Instructions pratiques sur l'observation et la mesure des propriétes optiques appelées rotatoires: avec l'exposé succinct de leur application à la chimie médicale, scientifique et industrielle* (Paris: Bachelier, 1845).

fulfilling the functions. Arago was a vocal denouncer of the practice. In 1836, the Académie Française offered Arago a place, drawing on the tradition of rewarding the secretaries of the Academy of Sciences for the literary efforts of their éloges.[3] Arago refused, with a very public denouncement in *le National* of " Académiciens doubles, triples, quadruples."[4] Biot, meanwhile, enjoyed the collection of positions, and ended up as one of the rare triple Academicians of France. Perhaps, however, he still smarted from the moment in 1848, when Arago and the Provisional Government abolished the practice, forcing Biot to give up at least one of his teaching positions.[5]

Thus even to the deathbed, Arago and Biot continued to fight over representation. How did one go about connecting signifier and signified? What was it that linked Arago to his title as perpetual secretary? Was it the pronouncement of authority, decreeing him as such? Or was it that Arago acted like the perpetual secretary, showing up to the Academy and fulfilling his duties? For Biot, it was the Academy's authority, this act of declaring him its secretary, that tied Arago to his position. He offered the salary as a symbolic bond affirming that Arago remained the secretary whether or not he acted the part. But Arago would have nothing to do with Biot's symbolic bonds, no matter how nicely offered. He had spent too much of his life fighting the arbitrary wielding of authority to be bought off by it now. For Arago, it was fulfilling the duties of the secretary of the Academy of Sciences that made him the secretary of the Academy of Sciences. He did not want the money, no matter what the Academy said. As Biot insisted, Arago remained firm, and their last interaction ended at an impasse. Even at deathbeds, sometimes reconciliation comes hard.

Biot's visit was one of the last Arago received. He died just a few days later on October 2, 1853. His funeral was "full of mud, rain, and regiments of soldiers."[6] Biot lived several more years, dying in 1862. Two years later, in 1864, both Arago and Biot would appear once again upon the topography of Paris. In the midst of renovating Paris, Baron Haussmann named a street after each of them. Boulevard Arago was a massive thoroughfare in the 13th, one of Haussmann's major spokes intended to open the city and render it legible. The Boulevard ran from the Observatory to the Latin Quarter, from Arago's home to his constituency, retracing the path taken by the working men of Paris when they marched to the Observatory to thank Arago for his support of their voting rights (and, cruel trick, assuring that the barricades they formed in 1848 could not be replicated).

Rue Biot is found on the opposite end of Paris. It is a single narrow block, cocked at a diagonal between the Place de Clichy and the Rue des Dames. This area was a noted red-light district on the outskirts of the city. Inhabited by prostitutes and other members of the demimonde, it was a place of obscurity

---

[3] Maurice Daumas, *Arago 1786–1853, la jeunesse de la science* (Paris: Belin, 1987), 191.
[4] *Le National*, April 28, 1836.
[5] Archives Nationales f17 20178 f. 85.
[6] J.B. Boucher, *Souvenir de jeunesse d'un Picard sur le clocher de son village* (Versailles: Impr. de Brunox, 1864) 2: 37.

and half-light. This was hardly the world Biot had frequented while alive, but its impenetrability captures something of Biot's own insistence on the continuation of mystery in the world.

Thus Paris inherited the dual legacy of its two leading optics researchers. Transparency and opacity continued to circle one another in a stalemate. The issue had divided them from the earliest moments of their work. From the first moments they each developed their polarimeter, Arago and Biot fought over what this instrument allowed one to see. The principal technical point in their argument over polarization was how to assign the color coefficient in the intensity equations of the polarimeter's results. Biot claimed that the colors were analogous to those of Newton's rings, and their true, complex nature could only be known by consulting the master list provided in Newton's *Opticks*. Arago dismissed this, maintaining the colors could be decided upon by universal agreement.

The equations for the colors of polarized light were taken by many nineteenth-century physicists as the model of a new discipline of physics that combined mathematical and experimental rigor. Yet at the heart of these equations, in the disputed color coefficient, were questions about representation that did not belong to mathematical physics alone. How direct was one's access to the natural world? Were there parts of the world that could not be represented at all? What implications did this have for the ability of people to come to agreement with one another?

The question of visibility is ultimately a question of optical instruments. Few people would be so strict as to claim that all we can know about the world is what happens to cross our field of vision by chance. If something is too far away for one to make it out distinctly, it is perfectly within the rules to walk closer to get a better look. The not seen is not necessarily the not seeable, and the world often happily complies with one's efforts to get a better look. By the beginning of the nineteenth century, these efforts had come to include a wide array of optical devices intended to bring the world into focus. Telescopes, microscopes, spyglasses, and spectacles all provided reliable representations of what existed in the world. The optical piece had been incorporated into an act of seeing that combined eye and artifact together into a single technological complex.

Instruments thus allowed one to rewrite the boundaries of what was visible and what was not. They demoted objects from unseeable to merely previously unseen. In doing so, they posed the category of unseeable itself as a question. Would the right instruments, correctly managed, allow us to see into every corner of the world? Or are there some pockets that would always remain beyond our grasp? This was the question before Arago and Biot as they began using and promoting the nineteenth century's most novel optical device: the polarimeter.

For Arago, the polarimeter was the saving grace of his regime of universal visibility. It would solve the nagging problem of individual differences that threatened the idea of a universal observer. He made several attempts to transform the instrument into a photometer that would give a quantitative reading of light intensity. This device would then make the observations of its various users interchangeable and would underpin the universally accessible science Arago envisioned as the director of the Paris Observatory.

For Biot, the polarimeter cemented the value of the unseen. He used his version of the device to test the optical activity of various substances, a characteristic that allowed him to distinguish between organized and unorganized bodies. Only materials that were or had once been living, he claimed, possessed the optical quality of rotating the plane of polarization of light. For Biot, using the instrument was not about seeing at all. The light emerging from his optics was not a representing image but a trace to the unrepresentable.

As their work continued, the implications stretched their way further and further into different domains of French culture. The abstract question of color perception became concrete in the polarimeter, an instrument in whose invention they shared even while they disagreed over what, precisely, it showed and who had the final word in the matter. Arago aimed his optical instrument at the heavens to demystify the light emanating from celestial bodies and drain them of unseen significance. Biot, on the other hand, maintained an opening for the possibility that celestial radiation may have an effect beyond the visible, and spent a good portion of his life reconstructing ancient systems making the same claim. The old Enlightenment debate over passive and active matter received a nineteenth-century update with the theory of radiations. The suggestion that one may be subject to a range of mysterious, undetectable forces undermined the case for free will. This problem of freedom conditioned Arago and Biot's political visions. For Arago, removing inequality would lay the foundation for a rational public. For Biot, a strong central authority was needed to counter the inevitable failure of transparent communication. Transparency connected the political and optical, not as mere metaphor, but ultimately as mechanism.

Index